Considerations on Novel Network Edge Physical Node Architectures

Leveraging Unprecedented (Bottom-up) Technology-centric Approaches Empowered by Micro/ Nanotechnologies (MEMS/NEMS) and Federated Learning, with an Outlook on AI/5G-empowered Applications

Published 2026 by River Publishers
River Publishers
Broagervej 10, 9260 Gistrup, Denmark
www.riverpublishers.com

Distributed exclusively by Routledge
605 Third Avenue, New York, NY 10017, USA
4 Park Square, Milton Park, Abingdon, Oxon OX14 4RN

Considerations on Novel Network Edge Physical Node Architectures
Leveraging Unprecedented (Bottom-up) Technology-centric Approaches Empowered by Micro/Nanotechnologies (MEMS/NEMS) and Federated Learning, with an Outlook on AI/5G-empowered Applications / Jacopo Iannacci, Vinod Belwanshi and Kandarpa Kumar Sarma.

Routledge is an imprint of the Taylor & Francis Group, an informa business

ISBN 978-87-4380-918-0 (paperback)
ISBN 978-87-4380-920-3 (online)
ISBN 978-8-743-80919-7 (ebook master)
ISBN 978-87-4380-922-7 (ePub)
DOI: 10.1201/9788743809197

Considerations on Novel Network Edge Physical Node Architectures

Leveraging Unprecedented (Bottom-up) Technology-centric Approaches Empowered by Micro/ Nanotechnologies (MEMS/NEMS) and Federated Learning, with an Outlook on AI/5G-empowered Applications

Editors

Jacopo Iannacci
Fondazione Bruno Kessler (FBK), Italy

Vinod Belwanshi
CSIR - National Metallurgical Laboratory, India

Kandarpa Kumar Sarma
Gauhati University, India

River Publishers

Routledge
Taylor & Francis Group
NEW YORK AND LONDON

Dedication

To all researchers, scholars, and innovators working tirelessly in the fields of semiconductors, communication technologies, and allied domains.

Your pursuit of knowledge, precision, and progress continues to illuminate the path toward a more connected, efficient, and intelligent world.

This work is dedicated to your unwavering commitment to advancing human know-how and shaping the technologies of tomorrow.

Dr. Kandarpa Kumar Sarma

Contents

Preface

There is no rest for the advancement of technology. Besides such a statement being applicable for a long time, it must be stressed that the pace at which such evolution progress is being made is getting faster and faster. Looking at the domain of telecommunications – which is the focus of this book – the long-lasting cycles, from conception to deployment of a new standard, which were measured in terms of decades, now develop in just a few years. In addition, the boundaries between the previous and next generations are becoming increasingly blurrier, making the transition change from an abrupt *switch ON* versus *switch OFF* less and less perceivable by the end-user. Given such a reference ground, this book copes with what it is expected to be, starting from 2030, the forthcoming complex paradigm of 6G (sixth generation of mobile communications), and on a longer time basis, of future networks (FNs).

It goes without saying that the scientific community has been active for years – to be precise from 2017–2018 – in developing a comprehensive discussion that will shape (and is already shaping today) 6G, along with numerous sets of use cases, key enabling technologies (KETs) and key performance indicators (KPIs). All this brings significant value to the evolution path that the standard will follow.

Just to mention a few relevant items that have been independently studied by various initiatives worldwide, and a preliminarily set for the upcoming standardization of 6G, in terms of KPI, a peak data rate of 1 Tbps and an average data rate of 10–100 Gbps per user will be reached, along with latency below 0.1 ms, and reliability (block error rate – BLER) better than 10^{-8}.

Turning now to 6G KETs and use cases, significant effort has been made on application domains such as, among other things, holographic communications, digital twins, multi-sensory extended reality (XR) gaming and entertainment, tactile/haptic communications, space-terrestrial integrated networks, and so on.

Such a complex scenario is already rich with expectations and challenges, and the goals in terms of enabled end-user services and functionalities seem to bear the promise of a better interconnected world. Yet, it is the belief of the authors of this book that something more should take place in order to turn 6G and especially FNs into reality, in the years and decades still to come.

In fact, if we look at the number of smart networked physical devices to be deployed and interconnected together across the infrastructure, estimates are simply impressive. To this end, the International Telecommunication Union (ITU) defined a set of challenging specifications to be targeted by the IMT-2030, i.e., International Mobile Telecommunications for 6G. Among such requirements, a number of physical devices as large as 10–100 Million per square kilometre is indicated for 6G. Of course, then, on a longer term basis, such a density indicator could do nothing but increase.

This said, and bearing the above scenario in mind, we believe that the leap to such a massive density of the network edge cannot be targeted simply by relying on the standard approaches and design philosophies that led – although successfully – to the development of co-designed HW–SW (hardware-software) systems up to this point. On a different perspective, we are convinced that unprecedented ways of conceiving the *physical atoms* that will be the pillars of edge intelligence (EI) within the context of 6G and FNs must be fostered and deployed.

By means of this book, we aim to inject new clues and ideas into the scientific discussion, capitalizing on physical (HW) technologies for manufacturing miniaturized sensors, actuators and transducers (SAT), namely, microtechnologies and nanotechnologies, as the building blocks of distributed intelligence of the future. We will also try to pursue the latter aim in conjunction with the involvement of the other half of KETs, i.e., non-physical technologies, like artificial intelligence (AI) and federated learning (FL), which we will leverage to discuss practical application scenarios.

Jacopo Iannacci, PhD

List of Contributors

Amrita Choudhury, *Research Scholar, Department of Electronics and Communication Engineering, Gauhati University, Guwahati, Assam, India.*

Aradhana Misra, *Assistant Professor, Department of Electronics and Communication Engineering, Gauhati University, Guwahati, Assam, India.*

Bhubon Chandra Mech, *Defence Institute of Advanced Technology (DIAT), Pune, India.*

Debasish Dev Misra, *Associate Professor, Department of Computer Science and Engineering, Faculty of Engineering and Technology, Assam down town University. Guwahati, Assam, India.*

Jacopo Iannacci, *Senior Researcher at Fondazione Bruno Kessler (FBK) Via Sommarive, Trento, Italy.*

Kandarpa Kumar Sarma, *Professor and Head, Department of Electronics and Communication Engineering, Gauhati University, Guwahati, Assam, India.*

Koushik Guha, *Associate Professor & Head, Dept. of Electronics & Communication Engineering, National Institute of Technology Silchar, Assam, India*

Lachit Dutta, *Assistant Professor, Department of Electronics and Communication Engineering, Gauhati University, Guwahati, Assam, India.*

Manasjyoti Bhuyan, *Research Scholar, Department of Electronics and Communication Engineering, Gauhati University, Guwahati, Assam, India.*

Praful Kabadi, *Defence Institute of Advanced Technology (DIAT), Pune, India.*

Purabi Sharma, *Research Scholar, Department of Electronics and Communication Engineering, Gauhati University, Guwahati, Assam, India.*

Rajesh Singh, *Defence Institute of Advanced Technology (DIAT), Pune, India.*

Vinod Belwanshi, *Scientist, CSIR- National Metallurgical Laboratory, Jamshedpur, India, (Ministry of Science and Technology, Government of India); Assistant Professor, Academy of Scientific & Innovative Research (AcSIR), Ghaziabad, UP, India; Honorary Research Fellow, University of Glasgow, UK G12 8QQ*

List of Abbreviations

4G	4th generation of mobile communications
5G	5th generation of mobile communications
6G	6th generation of mobile communications
AC	Affective Computing
AI	Artificial Intelligence
AlN	Aluminium Nitride
ANN	Artificial Neural Network
AP	Access point
ASIC	Application Specific Integrated Circuit
B5G	Beyond 5G
BCI	Brain-Computer Interactions
BS	Base Station
CIM	Computing In-Memory
CMOS	Complementary Metal-Oxide Semiconductor
CNN	Convolutional Neural Network
CRAS	Connected Robotics and Autonomous Systems
CSI	Channel State Information
DC	Direct Current
DCCS	Distributed Computing Continuum Systems
DL	Deep Learning
DLT	Distributed Ledger Technologies
DoF	Degree of Freedom
DRL	Deep Reinforcement Learning
E2E	End-To-End
EADO	Energy-Aware Distributed Optimization
ECG	Electrocardiography
EH	Energy Harvesting
EH-MEMS	MEMS for EH

EH-NEMS	NEMS for EH
EI	Edge Intelligence
ES	Enhanced Security
FEP	Free Energy Principle
FL	Federated Learning
FN	Future Network
GPS	Global Positioning System
HPC	High-Performance Computing
HW	Hardware
IC	Integrated circuit
ICEGs	Intracardiac electrograms
IMT	International mobile telecommunications
IMU	Inertial measurement unit
IoT	Internet of Things
ITU	International telecommunication union
KET	Key enabling technology
KPI	Key performance indicator
LIS	Large intelligent surfaces
LSB	Least significant bit
LSTM	Long short-term memory
MARL	Multi-agent reinforcement learning
MCM	Multi-chip module
MEMS	Microelectromechanical systems
MIMO	Multiple-input multiple-output
ML	Machine learning
mMIMO	Massive multiple-input multiple-output
mmWave	Millimetre waves
MO	Manifold optimization
MSB	Most significant bit
NEMS	Nanoelectromechanical systems
NTN	Non-terrestrial network
PAt	Power attenuator
PCI	Percutaneous coronary intervention
PCU	Passive control unit
PhS	Phase shifter
PLS	Physical layer security
PS	Paradigm shift
QKD	Quantum key distribution
QT	Quantum technologies
RF	Radio frequency
RFFE	Radio frequency front end

RF-MEMS	MEMS for RF applications
RF-NEMS	NEMS for RF applications
Rx	Reception
SAT	Sensors, actuators and transducers
SC	Seamless coverage
SD	Spectrum diversity
SI	Scattered intelligence
SOMP	Sparse orthogonal matching pursuit
SRIA	Strategic research and innovation agenda
SW	Software
SWIFT	Silicon wafer integrated fan-out technology
SwU	Switching unit
TEG	Thermoelectric generator
TI	Tactile internet
TRM	Transmission reception module
TSDSI	Telecommunications Standards Development Society, India
TS-HB	Two-stage hybrid beamforming
TSV	Through silicon via
Tx	Transmission
UAV	Unmanned air vehicle
UE	User equipment
ULP	Ultra-low power
VLC	Visible light communication
WL	Wafer level
WLP	Wafer level packaging
WL-SiP	Wafer level system in package
WPT	Wireless power transfer
XR	Extended reality

About the Editors

Jacopo Iannacci (Senior Member, IEEE) was born in Bologna, Italy, in 1977. He received his M.Sc. (Laurea) degree in electronics engineering from the University of Bologna, Italy, in 2003, and his Ph.D. in information and telecommunications technology from the University of Bologna, Italy, in 2007. He received Habilitation as Full Professor in Electronics from the Italian Ministry of University and Research (MUR), in 2021.

He worked in 2005 and 2006 as visiting researcher at the DIMES Technology Center (currently Else Kooi Lab) of the Technical University of Delft, the Netherlands, focusing on the development of innovative packaging and integration technology solutions for RF-MEMS devices. In 2016, he visited as seconded researcher the Fraunhofer Institute for Reliability and Microintegration IZM in Berlin, Germany, to conduct high-frequency characterization of RF-MEMS components jointly with the RF & Smart Sensor Systems Department at IZM. Since 2007, he has been a researcher (permanent staff) at the Center for Sensors & Devices of Fondazione Bruno Kessler, in Trento, Italy.

His research interests and experience fall in the areas of finite element method multi-physics modelling, compact (analytical) modeling, design, optimization, integration, packaging, experimental characterization and testing for reliability of MEMS and RF-MEMS devices and networks for sensors and actuators, energy harvesting (EH-MEMS) and telecommunication systems, with applications in the fields of 5G, Internet of Things, as well as 6G, tactile internet and Super-IoT.

Dr. Iannacci has authored more than 180 scientific contributions, including international journal papers, conference proceedings, books, book chapters and one patent. He is Associate Editor of Springer Microsystem Technologies,

Series Editor of the IEEE Communications Magazine, and Distinguished Lecturer of the IEEE Communications Society 2025 2026.

Vinod Belwanshi is a Scientist in the Materials Engineering Division at CSIR-National Metallurgical Laboratory (CSIR-NML), Jamshedpur, JH, India. He also holds an Honorary Research Fellow position at the University of Glasgow, UK, and is recognized as an Assistant Professor at the Academy of Scientific and Innovative Research (AcSIR), Ghaziabad, UP, India.

He received his M.Tech. in Instrumentation from the Indian Institute of Technology, Kharagpur, India. He was awarded the prestigious Department of Atomic Energy Graduate Fellowship Scheme (DGFS) to pursue his Ph.D. in Engineering Sciences from Bhabha Atomic Research Centre, Mumbai, India, under the aegis of the Homi Bhabha National Institute, Mumbai, India. Dr. Belwanshi is also a recipient of the prestigious Ramanujan Fellowship from the Department of Science and Technology, Government of India, aimed at attracting exceptional Indian researchers working abroad to return to India.

Dr. Belwanshi worked as a Senior Project Research Scientist and Visiting Researcher at the Indian Institute of Technology Bombay, Mumbai, India, where he focused on memory devices and, micro/nanoelectronic devices based on SU-8 and carbon nanomaterials. He has served as a Research Associate at the University of Glasgow, UK where he worked on MEMS gravimeters. His current research focuses on non-destructive evaluation and testing and micro/nano electromechanical systems technologies for harsh environments generally present in nuclear, aerospace, and industrial sectors. He has published his research widely in reputed journals and conferences. He is a life member of the Institute of Smart Structures and Systems (ISSS) and the Indian Society for Non-destructive Testing (ISNT), and a member of the IEEE Sensors Council, and IEEE MEMS Technical Committee.

Kandarpa Kumar Sarma, currently Professor, Department of Electronics and Communication Engineering, Gauhati University, Guwahati, Assam, India specializes in mobile communication, artificial intelligence, human–computer interaction, cognitive and software defined radio, speech processing and antenna design. He earned his Ph.D. in mobile communication (2012) from IIT Guwahati. Furthermore, he completed PostDoc research in the area of application of AI in mobile communication from Technical University Sofia, Bulgaria.

Currently he is associated with research in the areas of pandemic compliant smart infrastructure development, AI aided approaches in dense wireless networks, cognitive spectrum management, electronic warfare and high data rate micro-strip patch antenna designs.

He has authored and edited twelve books. He has been a TPC member of over two hundred international conferences. He is a Senior Member of IEEE (USA) and a fellow of Institution of Electronics and Telecommunication Engineers (India). He serves as the Editor in Chief, International Journal of Intelligent Systems Design and Computing, Inderscience, and Associate Editor, Network: Computation in Neural Systems.

Introduction

Looking at 2030 and beyond, the forthcoming complex paradigms of 6G (sixth generation of mobile communications), future networks (FNs) and Internet of Everything (IoE) will bring transformative disruption at different levels, starting from end-user applications and services to the way physical infrastructures will be conceived and operated.

Within such a multi-faceted scenario, this book turns attention on the network edge, stressing how the forecasted proliferation and technology innovation of the physical systems at the boundary of the network (edge), will pose crucial issues to be addressed in the next few years. In particular, the continuity from the edge to the core of the network will be addressed, suggesting unprecedented design and development approaches (Chapter 1).

Along such a direction, and embracing an unprecedented bottom-up approach, the potential impact of micro and nanotechnologies as a key enabling technology (KET) for the network edge of the future, along with so-called edge intelligence (EI), is sketched. This includes numerous examples of already existing micro/nano (MEMS/NEMS) devices, components and systems, implementing a variety of sensors, actuators and transducers (SATs) (Chapter 2).

Still in the context of miniaturized SATs, fabrication processes adapted from the integrated circuit (IC) industry, along with specialized techniques designed for M/NEMS devices are covered. Along such a direction, an in-depth look at various fabrication steps employed in the development of devices intended for FN is also provided (Chapter 3).

Then, a practical study of how improving hardware technologies at the edge can be beneficial in terms of more efficient operation is reported. To this end, scattered intelligence shall be a key enabler with federated learning (FL) at the base station providing decisive assistance to adaptive downlink beamforming codebook design for mm-wave (millimetre wave) massive MIMO (multiple-input-multiple-output) set-ups as essential ingredients for enhanced link reliability in edge networks (Chapter 4).

Completing the book with a perspective on distributed applications, the monitoring of cardiovascular diseases capitalizing on artificial intelligence (AI) and 5G is also covered (Chapter 5).

In the following, titles and abstracts of each chapter featured in the book are listed.

Chapter 1: Considerations on Transformative Needs at the Edge of 6G and of Future Networks

This opening chapter aims at providing a common understanding and reference ground of 6G and FNs, useful for building further considerations on how the infrastructure's edge will evolve. Given its complexity, the discussion of 6G is reconducted to four main paradigm shifts (PSs) that will bring innovation and disruption. They are linked to: (a) extensive exploitation of artificial intelligence (AI) (both on user and network ends); (b) infrastructure diversification ensuring ubiquity of key performance indicators (KPIs); (c) exploitation of diversified frequency spectra; (d) advanced solutions to ensure privacy, security and trust of data. Thereafter, considerations on the transformative trends that will drive the evolution of the 6G/FN (sixth generation of mobile communications/ future networks) network at the edge are elaborated. In this sense, particular attention is devoted to the huge proliferation of edge devices, which will require unprecedented characteristics of the nodes, such as, e.g., cognitive-like features, self-management and self-recovery capacities, along with local storage and elaboration of data. Driven by such an augmented independence of each network segment, the aspect of orchestration and of its complexity will be covered, aiming at keeping the network of networks work as a whole. Finally, the current status of 6G development, concerning developed use cases as enabling technologies, is reported, including building a comparison with what was envisioned by the scientific community in the early days of discussion, around 2018–2019.

Chapter 2: Micro and Nanotechnologies as a Key Enabling Technology of a Future Network Edge

This chapter develops a comprehensive discussion on the characteristics and features that low-complexity hardware (HW) components, i.e., sensors, actuators and transducers (SATs), will need to be implemented in order to enable the network edge infrastructure of 6G and FN. To this end, micro and nanotechnologies (MEMS/NEMS) are identified as key enabling technologies (KETs), capitalizing on their typically high miniaturization, integration and functional diversification, the latter is enabled by the substantial inexpensiveness of HW redundancy. In particular, a set of ten founding technology and functional trends are identified. Such items will drive the architecting and evolution of the 6G/FN network edge, and all of them are notably met by micro/nanotechnologies. In brief, the set of ten technology trends is related to: (1) resilient self-adaptation of operation and functioning of SATs; broadening and diversification of (2) sensing, (3) actuating, and (4) transducing functionalities; (5) increase in miniaturization and integration of basic physical SAT components; (6) provision, storage and distribution of energy to enable autonomous operation of edge physical nodes; (7) evolution of functions to cope with altered and unexpected operation conditions; (8) transmission and reception of data, keeping network edge physical nodes linked to the core and orchestrated along with the rest of the infrastructure; and, finally, (9) data storage and (10) computational capacities, both resident within edge physical nodes, increasing the locally implemented functions and their autonomy with respect to the whole network.

Chapter 3: Micro/Nano Fabrication Processes – Utilizing Borrowed and Dedicated Fabrication Processes for 6G and Future Network Devices

Micro/nanofabrication techniques and processes play a vital role as they are crucial in enabling miniaturised and light weight compact devices and systems. The ability to have precise and reliable fabrication with sub-10 nm features is one of the key requirements for driving innovation and advancement in nanoscience and nanotechnology, and it is also indispensable for many non-CMOS devices.

While conventional lithography is a mainstream approach, it is limited in resolution and precision for realising the micro/nano devices that are being targeted for use in the 6G future network. Unconventional methods like mechanics-enabled and post-trimming provide complementary advantages by enabling control in microstructure geometry, material adaptability and precision tuning beyond the capabilities of conventional lithographic methods.

These unconventional methods are suitable for low temperature applications. The post trimming techniques includes focused ion beam (FIB) milling, atomic layer etching, thermal or chemical trimming and electrochemical tuning.

The chapter explores fabrication methods for advanced micro/nano devices, including synthesis, on-chip transfer, film patterning, and property tuning. It also explores the fabrication methodology of dielectric meta surfaces using standard nanolithography techniques and extreme nanofabrication, highlighting methods like EBL, FIB, laser-based nanomanufacturing, grayscale lithography, multistep lithography, nanoimprint lithography, deep reactive ion etching and many other techniques that are borrowed from IC technology. The application of 3D printing for electromagnetic metamaterial absorbers is also discussed.

The context of these fabrication innovations includes potential applications in 6G wireless networks, particularly terahertz (THz) communications, where reconfigurable intelligent surfaces (RIS) and metasurfaces are proposed to address challenges like signal loss. Fabrication is also critical for emerging applications such as quantum devices, nano-optics, and high-performance sensing, often requiring sub-10 nm features and novel materials.

The chapter will also touch upon MEMS-based devices and their integration, and the challenges associated with these advanced fabrication techniques, such as scalability, throughput, and their cost effectiveness.

Chapter 4: Federated Multi-agent Reinforcement Learning for Edge-intelligent Beamforming Codebook Design for mmWave MIMO

Hybrid beamforming with massive multiple-input-multiple-output (MIMO) antenna arrays plays a pivotal role in achieving high data rates and spectral efficiency in next-generation wireless systems. Nevertheless, the real-time design and adaptation of beamforming codebooks and user grouping remain challenging, as conventional machine learning (ML) methods typically rely on the centralized collection of extensive channel state information (CSI), resulting in substantial feedback overhead and potential privacy violations.

This chapter proposes a federated multi-agent reinforcement learning (MARL) framework for joint hybrid beamforming and dynamic user grouping. Unlike prior works which focus solely on centralized codebook learning, the proposed approach enables distributed user equipment (UE) to collaboratively train codebook and clustering algorithms without sharing raw data, thereby preserving privacy.

The proposed method significantly reduces communication overhead while achieving near-optimal beam patterns and robust scalability in dense network scenarios, making it well-suited for practical deployment in next-generation wireless systems. The experimental findings indicate that for mobile users, federated learning (FL)-based user clustering with MARL outperforms traditional initial access-based beam clustering in MARL.

Chapter 5: Advancements in Cardiac Monitoring: The Role of AI and 5G Networks in Optimizing Cardiovascular Healthcare

Cardiovascular diseases have remained one of the major health burdens worldwide; their timely detection and thus accurate diagnosis have become key elements in optimizing patient outcomes. In this chapter, we review artificial intelligence applied with 5G networks for cardiac monitoring. While artificial intelligence and data analytics have improved diagnostic accuracy and treatment planning for invasive techniques, non-invasive approaches, such as wearable devices and remote monitoring, have been increasingly exhibiting accuracy in the early identification of potential problems. However, cardiac monitoring with the use of AI and 5G technology is yet to overcome many challenges. These include a host of issues concerning high-quality data acquisition on a large scale, the much-needed standardization and interoperability among disparate devices and platforms, privacy and security issues, and finally, the cost-effectiveness of the proposed system. Further research and development in AI and 5G technology will likely bring enormous change to coronary healthcare and results for patients.

CHAPTER

1

Considerations on Transformative Needs at the Edge of 6G and of Future Networks

Jacopo Iannacci

Fondazione Bruno Kessler – FBK, Via Sommarive, 18. Trento – 38123, Italy

Email: iannacci@fbk.eu

Abstract

This opening chapter aims at providing a common understanding and reference ground for 6G and FN, useful for building further considerations on how the infrastructure's edge will evolve. Given its complexity, the discussion of 6G is structured according to four main paradigm shifts that will bring innovation and disruption. They are linked to (a) extensive exploitation of AI (both on user and network ends); (b) infrastructure diversification ensuring ubiquity of KPIs; (c) exploitation of diversified frequency spectra; (d) advanced solutions to ensure privacy, security and trust of data. Thereafter, considerations on the transformative trends that will drive evolution of the 6G/FN network at the edge are elaborated. In this sense, particular attention is devoted to the huge proliferation of edge devices, which will require for unprecedented characteristics of the nodes, such as, e.g., cognitive-like features, self-management and self-recovery capacities, along with local storage and elaboration of data. Driven by such an augmented independence of each network segment, the aspect of orchestration and of its complexity

DOI: 10.1201/9788743809197-1

will be covered, aiming at keeping the *network of networks* work as a whole. Finally, the current status of 6G development, concerning developed use cases as enabling technologies, is reported, also building a comparison with what was envisioned by the scientific community in the early days of discussion, around 2018–2019.

1.1 Introduction

As of now, the discussion mainly inspired in recent years by 6G, is sketching sets of diversified services and applications, KPI, and KET, which are so challenging that they seem to be falling well-beyond the targets of the 6G standard itself. [1]. In fact, observing the way other generations (including 5G) evolved, it is as though 6G is going to draw a line between the previous telecommunications-centric networks, and the upcoming services- and applications-centric diversified and distributed infrastructures.

In practical terms, 6G can be briefly described as a bearer of innovation over two distinct planes. On one hand, in full continuity with previous generations, critical KPIs, such as, among others, data rates available per user, latency and reliability, are expected to improve between 100 and 1000 times with respect to what 5G will mark in its lifespan. On the other hand, AI will be leveraged at each level of the network infrastructure, both to improve services (end user plane), as is already ongoing today, as well as, in disruption with the past, to manage more efficiently the infrastructure itself (network operation plane). This means that the 6G network is envisaged to work as a *network of networks*, with augmented self-management capacities of each segment of the infrastructure and pronounced orchestration, to continue working as a whole.

Given such a diverse context, it is straightforward that the complexity at different levels, spanning from technologies to algorithms, is going to be enormous.

1.2 Paradigm Shifts Driven by 6G

As already mentioned, trying to capture most of the forecasted disruption of 6G within a brief narrative is a non-trivial task. A brilliant approach is developed in [2], where KPIs and KETs are grouped according to a small number of PSs that will be triggered by 6G. The four PSs, after being re-termed and rearranged in [3],[4], are graphically shown in Figure 1.1.

Figure 1.1 Graphical arrangement of the four PSs of 6G as re-termed in [3],[4]. (Images and thumbnails composing the graphic are powered by https://www.Freepik.com.)

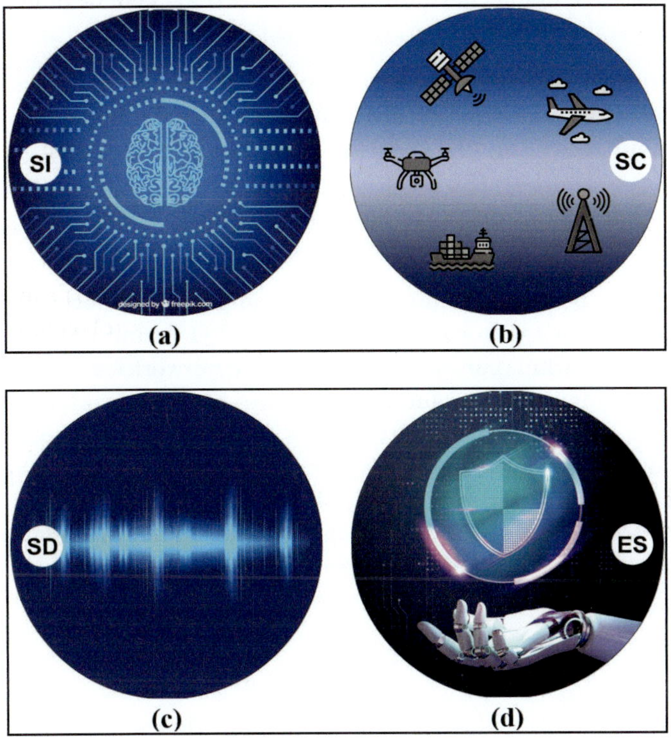

In a nutshell, the four PSs can be briefly recapped as follows:

- **Scattered intelligence (SI)**. Exploitation of AI at each level of the network infrastructure
- **Seamless coverage (SC)**. Diversified infrastructure ensuring the same KPI everywhere
- **Spectrum diversity (SD)**. Diversification of frequency spectra supporting increased data volume
- **Enhanced security (ES)**. Improving privacy, security and trust of data scattered everywhere

More extended discussion will be developed in the following sections.

1.2.1 Scattered intelligence (SI)

This PS hinges around the cornerstone technology of AI (see Figure 1.1a). Since the very early discussion of 6G, the backbone of decentralized data and computation across the network has been stressed as a crucial trend to be addressed [5],[6]. Further elaboration of such a proposition led to forecasting a massive exploitation of AI at each level of 6G infrastructure, with reference both to the service plane (in continuity with previous generations), and (more disruptively) to the network operation plane [7],[8].

AI algorithms can, e.g., optimize network operation by addressing problems that involve large amounts of diversified data, including mining, sensing, prediction and reasoning [9]. Focusing on the user-end, AI can put together the sensing of network traffic variations, of resources utilization, averaged user demands, along with identification of potential threats, optimizing on a real-time basis the functioning of various network entities, like BS and UE [10]–[12].

Also, it must be borne in mind that emerging wireless services will evolve into complex systems with varying requirements [13], and the characteristics of AI and ML algorithms can empower self-aware networks. Further ahead, this can lead to developing self-evolutionary features of the network of networks (or system of systems), with the ability, among others, to gather proper resources to ensure local maintenance of KPIs (i.e., elevated resilience), while keeping homogeneity with the network as a whole. This implies complex orchestration of the infrastructure [14] as well as, on the other hand, distribution of AI at each level of the network, which (re-termed) is EI [15].

1.2.2 Seamless coverage (SC)

This PS builds around another fundamental proposition of 6G, which is ensuring the same KPIs, regardless of whether the user (human or machine) is accessing the network from a metropolitan, rural or remote location. Such a remarkable target is driven by the mega-trend towards ubiquity of services, and is already being pursued under 5G through extensive diversification of the physical infrastructure [16],[17]. This concept is shown schematically in Figure 1.1b, where the network relies on terrestrial, airborne, space and oversea nodes. A comprehensive report on this topic is available in [18],[19]. Given such a frame of reference, the topic of NTNs is being intensively discussed recently [20], with particular focus on aspects at different levels, arising from integration of satellites within the classical ground infrastructure [21].

1.2.3 Spectrum diversity (SD)

Recalling a few crucial expected KPIs of 6G, like breaking the 1 Tbps average data rate per user, scoring E2E latency below 1 ms, and ensuring E2E reliability of 99.99999% [22], it is easy to guess that non-conventional exploitation of the frequency spectrum will be necessary, well beyond what 5G is already pursuing today (see Figure 1.1c).

That said, the discussion in [2] forecasts for 6G an all spectra scenario, leveraging the continuity across sub-6 GHz frequencies, mmWave (10–30 GHz), sub-THz (30–300 GHz), THz range (above 300 GHz), along with optical frequencies, including the visible light spectrum (e.g., VLC). Such a variety of spectra gives rise to a plethora of pros and cons that must be carefully and thoroughly evaluated. To this end, comprehensive discussion of various related aspects, ranging from the physics of data transmission at such high frequencies to key-features as advanced beamforming, and spectrum policy issues, is available in [23]–[26].

1.2.4 Enhanced security (ES)

Having in mind the discussion above, it is straightforward that 6G-driven high integration of technologies and ubiquity of services, along with massive exploitation of AI, will raise so many and non-trivial issues in terms of security, privacy and trust of data, that a dedicated PS (i.e., ES) is necessary for their counteraction and mitigation [27]–[29] (see Figure 1.1d).

The growth in sensing functionalities and the prevalence of mobile/wearable devices centred around human-centric services and communications, will prompt security/privacy challenges across multiple fronts. AI itself presents several issues, encompassing data security, AI model and algorithm security, SW vulnerabilities, and the improper utilization of AI technologies. The training of AI models involves collecting vast amounts of data, likely containing users' sensitive information such as identity and location [2].

The proliferation of miniaturized and wearable IoT devices raises security concerns due to frequent interactions and interconnections, emphasizing the need for efficient authentication mechanisms and strategies. Traditional encryption/decryption techniques become cumbersome in this context, as they demand computing resources that are often unavailable given the limited computational and storage capability and power availability typical of IoT devices. Similar considerations apply to UAV networks (recalling the PS of SC), prompting the development of lightweight encryption protocols [30],[31].

Just to list (in no particular order) a few key technologies and strategies that can provide significant contribution in the PS of ES, the following ones are worth to be mentioned:

(1) Acting at the PLS level [32]
(2) Exploitation of QT, e.g., through QKD [33],[34]

(3) Blockchain [35],[36]
(4) FL [37],[38].

1.3 General Considerations on the Future Network Edge and Integration with the Whole Infrastructure

This section develops some considerations on the future development of the network edge from the perspective of physical systems. To do so, the features expected for the 6G/FN network edge are first briefly discussed. Then, the resulting open issue of orchestrating an increasingly more complex edge with the rest of the infrastructure is introduced.

1.3.1 Proliferation and disruption at the edge of the network

Recalling the discussion developed to this point, it is out of the question that the edge of the network will relentlessly proliferate in the future, driven by application paradigms like the (Super-)IoT and the TI, with 6G and FN covering connectivity to the rest of the network (i.e., towards the core) and of the world. This statement comes with important implications to be addressed. In particular, two considerations should be borne in mind.

The first, in close continuity with what is already ongoing, is that the density of devices connected to the network per square kilometre of 0.1 M/km^2 under 4G, is expected to rise to 1 M/km^2 with 5G, and to 10 M/km^2 when 6G takes over [39] (see Figure 1.2).

Figure 1.2 Millions of devices per square kilometre from 4G to 6G [39].

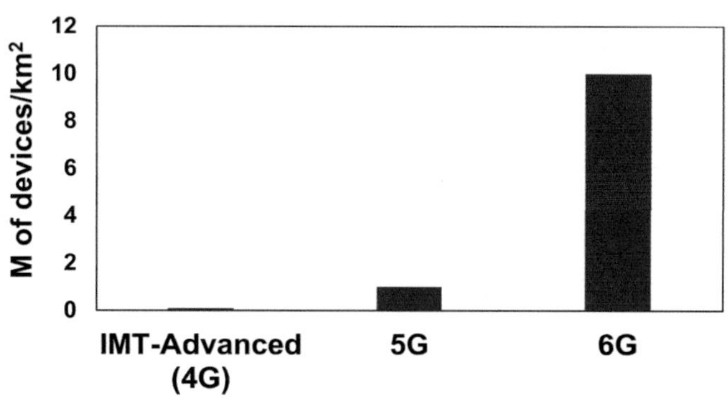

The second is linked to the inherent disruption of 6G and FN (as sketched in Sections 1.1 and 0). This has to do with conceiving and designing edge physical systems that implement bio-/nature-inspired concepts, which is far from how semiconductor-based electronics and telecommunications evolved across the past five to six decades. Just to mention a few of these unprecedented characteristics, the following items must be reported:

(1) Cognitive-like features, e.g., self-reaction and self-repair intrinsic capacities [40],[41].
(2) Self-management, high resilience and self-evolution of physical systems [42],[43].
(3) Energy autonomy and self-sustenance (zero-energy systems and EH) [44],[45].
(4) Post-digital and non-conventional computing (also termed soft-computing) and CIM systems [46],[47].
(5) AC [48],[49].

Now that the crucial emerging trends of capillarization and disruption of the network edge are addressed, the consequent aspect of integration within the whole infrastructure will be addressed.

1.3.2 Addressing continuity from the edge to the core

Even at a brief glance, making the FN edge work as a whole with the rest of the infrastructure is not trivial. This is known to the scientific community, and efforts are already being spent to address such a complex issue. The resulting research is not yet aggregated in an established discipline, but it can effectively be termed as a DCCS [50].

In fact, already today, systems that leverage the Cloud infrastructure along with new computing tiers, like Fog and Edge, are working as a computing continuum [51]. However, the discussion in [52] lists a certain number of potential weaknesses, along with boundary conditions, which may yield that such classical computing continuum approaches are not the best option for the network edge of the future.

First, existing strategies are devoted to solving specific challenges. Moreover, they rely on traditional top-down methods for design and management, deriving from the first Internet-based systems, i.e., with a server and a client, specified by the SW application.

Also, [52] points out that edge entities may not count on a certain architecture. Therefore, the architecture itself should become a dynamic feature of the systems, triggering the ability of reacting to (unexpected) situations. By contrast, in the classical approach, the system characteristics are driven by their infrastructure. To this end, a key sentence of [53] is quoted in [52], i.e., "a system is complex if its behaviour crucially depends on the details of the system".

Another important statement, reported in [52] while discussing the novel approach proposed, is that a DCCS addresses all the resources required to enable an application of the computing continuum; this also includes the application itself but just as another component of the system.

In addition, it is argued that in the Cloud a certain system infrastructure can have various characteristics, counting on the available DoF in terms of resources to be employed. However, as the infrastructure starts approaching the Edge, the resources are increasingly limited, the DoF narrower, and the infrastructure approaches the use of all its capabilities.

The work in [52] also covers relevant principles, like that of reactive systems, of equilibrium and of FEP, borrowed from neuroscience, with reference to the brain adaptive behaviour [54]. It must be noted that all the considerations in [52] mentioned above, refer to systems based on SW and to their management through a novel approach based on Markov chains [55]. As the focus of this work is that of micro and nanotechnologies as potential KETs of FN edge physical systems, it seems that no overlaps exist.

In fact, the propositions in [52] stressed above, despite referring to a different field of research, carry twofold relevance for the discussion at stake here. On the one hand, it is clear that the classical approaches, still in use today, will not be effective in managing the complexity of the future edge.

From a different perspective, collective requirements and constraints are identified, like the need for high resilience, self-management and self-evolution, along with the scarce availability of resources, among which, sophisticated HW items and energy.

1.4 Current Stage of 6G

As last section of the chapter, it is worth taking a snapshot of the current stage of development of 6G, focusing on KETs and applications use-cases. To this end, a comparison deferred in time will be elaborated between one of the first relevant perspective contributions on 6G, reported in [22] and published in 2020, and a very recent paper discussing technology trends, available in [56] and published in late 2024.

This said, the set of driving applications and KETs of 6G predicted in [22] develops as follows:

- Multisensory XR applications
- CRAS
- BCI
- Blockchain and DLT
- Above 6 GHz for 6G
- Transceivers with integrated frequency bands
- Communication with LIS
- Edge AI
- Integrated terrestrial, airborne, and satellite networks
- Energy transfer and EH.

Concerning [56], relevant activities on 6G are summarized in Table 1.1, where a list of key use cases are cross-referenced with the initiatives worldwide addressing them.

Table 1.1 Mapping of 6G use cases and initiatives worldwide developing them, as reported in REF. Information on items labelled "A" to "H" are listed in Table 1.2.

6G use cases	A	B	C	D	E	F	G	H
Holographic communications	X	X	X	X	X	X	X	X
Cyber-physical systems, digital twins, manufacturing	X	X	X	X	X	X	X	X
Multi-sensory XR, gaming, entertainment	X	X	X	X	X	X	X	X
Tactile/haptic communications	X	X	X	X	X	X		X
Medical/health vertical, telesurgery	X	X	X	X	X	X	X	
Cooperative operation among a group of service robots/ drones	X	X	X	X	X		X	X
Imaging and sensing	X	X	X	X	X			X
Transportation vertical (automotive, logistics, aerial, marine, etc.)	X	X	X	X	X		X	
Space-terrestrial integrated network	X	X		X	X		X	X
Intelligent operation network	X		X		X		X	X
Critical infra, government/national security	X	X		X				
First responder/emergency services		X		X	X			
Smart buildings			X	X	X			
Agriculture/smart farming				X	X			

The details of the initiatives labelled with letters from "A" to "H" in Table 1.1 are reported in Table 1.2.

Table 1.2 Details of 6G initiatives labelled "A" to "H" listed in Table 1.1.

Label	Initiative/Entity	Country	Ref.
A	Networld Europe SRIA 2022	Europe	[57]
B	5G Americas/Next G Alliance	North America	[58],[59]
C	Huawei	China	[60]
D	B5G Consortium	Japan	[61]
E	TSDSI	India	[62],[63]
F	MediaTek	Taiwan	[64]
G	Survey Paper	N/A	[65]
H	ITU IMT-2030	N/A	[66]

Eventually, comparing the list of items reported above as in [22] with Table 1.1 and [56], it is clear that most part of the KETs and pivotal applications forecasted in the early days of the 6G discussion are being further developed and are likely to be accounted for in the standardization phase that is going to be undertaken.

1.5 Conclusion

This chapter has sketched the reference ground for the subsequent discussion that will be developed across the next chapters. In particular, attention is given to three main, i.e., describing 6G as a whole, stressing transformative aspects at the edge of 6G/FN, and providing a snapshot of the current state of research and development.

First, given the inherent complexity of 6G, the discussion has been structured around four principal PS expected to drive innovation and disruption: (a) the pervasive integration of AI at both user and network levels; (b) the infrastructure diversification to ensure ubiquitous maintenance of KPIs; (c) the utilization of a broad range of frequency spectra; (d) the implementation of advanced strategies to guarantee data privacy, security, and trust.

Building on such pillars, the chapter has explored the transformative trends shaping the future of the 6G/FN network, with a focus on the massive growth of edge devices. This proliferation is expected to impose new demands on edge nodes, including cognitive capabilities, self-management and self-recovery functions, and support for local data processing and storage. As each segment of the network gains greater autonomy, the complexity of orchestration becomes a key concern, ensuring that this increasingly distributed *network of networks* continues to operate as a unique system.

Finally, the current state of 6G development has been reviewed, highlighting both the enabling technologies and emerging use cases. This assessment has been contextualized through a comparison with the initial visions proposed by the scientific community during the early stages of discourse in 2018–2019.

References

[1] C. -X. Wang et al., “On the Road to 6G: Visions, Requirements, Key Technologies, and Testbeds,” IEEE Communications Surveys & Tutorials, vol. 25, no. 2, pp. 905–974, Secondquarter 2023, doi: 10.1109/COMST.2023.3249835

[2] X. You et al., “Towards 6G wireless communication networks: Vision, enabling technologies, and new paradigm shifts,” Sci. China Inf. Sci., 2021, 64, 1–74. https://doi.org/10.1007/s11432-020-2955-6

[3] J. Iannacci and H. V. Poor, “Review and Perspectives of Micro/Nano Technologies as Key-Enablers of 6G,” in IEEE Access, vol. 10, pp. 55428–55458, 2022, doi: 10.1109/ACCESS.2022.3176348

[4] J. Iannacci, “A perspective vision of micro/nano systems and technologies as enablers of 6g, super-iot, and tactile internet [point of view],” in Proceedings of the IEEE, vol. 111, no. 1, pp. 5–18, Jan. 2023, doi: 10.1109/JPROC.2022.3223791

[5] H. B. McMahan, E. Moore, D. Ramage, S. Hampson, and B. Agüera y Arcas, “Communication-efficient learning of deep networks from decentralized data,” in Proc. 20th Int. Conf. Artif. Intell. Statist. (AISTATS), 2017, pp. 1273–1282

[6] M. Katz, M. Matinmikko-Blue, and M. Latva-Aho, “6Genesis flagship program: Building the bridges towards 6G-enabled wireless smart society and ecosystem,” in Proc. 10th IEEE Latin-Amer. Conf. Commun. (LATINCOM), 2018, pp. 1–9. doi: 10.1109/LATINCOM.2018.8613209

[7] L. Zhang, Y. C. Liang, and D. Niyato, “6G visions: Mobile ultra-broadband, super internet-of-things, and artificial intelligence,” China Commun., vol. 16, pp. 1–14, 2019. doi: 10.23919/JCC.2019.08.001

[8] K. B. Letaief, W Chen, Y Shi, J. Zhang, and Y. J. A. Zhang, “The roadmap to 6G: AI empowered wireless networks,” IEEE Commun. Mag., vol. 57, pp. 84–90, 2019. doi: 10.1109/MCOM.2019.1900271

[9] R. Li et al., “Intelligent 5G: When cellular networks meet artificial intelligence,” IEEE Wirel. Commun., vol. 24, pp. 175–183, 2017. doi: 10.1109/MWC.2017.1600304WC

[10] A. Bandi, “A review towards AI empowered 6G communication requirements, applications, and technologies in mobile edge computing,”

in Proc. Int. Conf. Comput., Commun. Multimedia (ICCMC), 2022, pp. 1–10. doi: 10.1109/ICCMC53470.2022.9754049

[11] A. Giannopoulos et al., “Supporting intelligence in disaggregated open radio access networks: Architectural principles, AI/ML workflow, and use cases,” IEEE Access, vol. 10, pp. 41973–41988, 2022. doi: 10.1109/ACCESS.2022.3166160

[12] W. Wu et al., “AI-native network slicing for 6G networks,” IEEE Wirel. Commun., vol. 29, pp. 42–49, 2022. doi: 10.1109/MWC.001.2100338

[13] M. G. Kibria et al., “Big data analytics, machine learning, and artificial intelligence in next-generation wireless networks,” IEEE Access, vol. 6, pp. 32328–32338, 2018. doi: 10.1109/ACCESS.2018.2837692

[14] M. A. Habibi et al., “The architectural design of service management and orchestration in 6G communication systems,” in Proc. IEEE INFOCOM Workshops, 2023, pp. 1–10. doi: 10.1109/INFOCOMWKSHPS57453.2023.10225846

[15] S. Kitanov and V. Nikolikj, “The role of edge artificial intelligence in 6G networks,” in Proc. Int. Conf. Inf. Commun. Technol. (ICEST), 2022, pp. 1–7. doi: 10.1109/ICEST55168.2022.9828567

[16] L. Boero et al., “Satellite networking integration in the 5G ecosystem: Research trends and open challenges,” IEEE Netw., vol. 32, pp. 38–45, 2018. doi: 10.1109/MNET.2018.1800052

[17] S. A. R. Naqvi et al., “Drone-aided communication as a key enabler for 5G and resilient public safety networks,” IEEE Commun. Mag., vol. 56, pp. 36–42, 2018. doi: 10.1109/MCOM.2017.1700451

[18] P. P. Ray, “A review on 6G for space-air-ground integrated network: Key enablers, open challenges, and future direction,” J. King Saud Univ. Comput. Inf. Sci., vol. 33, pp. 1–15, 2021. doi: 10.1016/j.jksuci.2021.08.014

[19] H. Cui et al., “Space-air-ground integrated network (SAGIN) for 6G: Requirements, architecture, and challenges,” China Commun., vol. 19, pp. 116–133, 2022. doi: 10.23919/JCC.2022.02.008

[20] M. M. Azari et al., “Evolution of non-terrestrial networks from 5G to 6G: A survey,” IEEE Commun. Surv. Tutor., vol. 24, pp. 1722–1759, 2022. doi: 10.1109/COMST.2022.3199901

[21] L. Han et al., “Evolution to 6G for satellite NTN integration: From networking perspective,” in Proc. IEEE Int. Conf. Netw. Protocols (ICNP), Reykjavik, Iceland, 2023, pp. 1–6. doi: 10.1109/ICNP59255.2023.10355589

[22] W. Saad, M. Bennis, and M. Chen, “A vision of 6G wireless systems: Applications, trends, technologies, and open research problems,” IEEE Netw., vol. 34, pp. 24–30, 2020. doi: 10.1109/MNET.001.1900287

[23] T. S. Rappaport et al., “Wireless communications and applications above 100 GHz: Opportunities and challenges for 6G and beyond,” IEEE Access, vol. 7, pp. 102978–102994, 2019. doi: 10.1109/ACCESS.2019.2921522

[24] A. Shafie et al., "Terahertz communications for 6G and beyond wireless networks: Challenges, key advancements, and opportunities," IEEE Netw., vol. 37, pp. 70–76, 2023. doi: 10.1109/MNET.118.2200057

[25] J. M. Jornet, E. W. Knightly, and D. M. Mittleman, "Wireless communications sensing and security above 100 GHz," Nat. Commun., vol. 14, pp. 1–13, 2023. [Online]. Available: https://doi.org/10.1038/s41467-023-36621-x

[26] M. J. Marcus, "6G spectrum policy issues above 100 GHz," IEEE Wireless Commun., vol. 28, pp. 7–8, 2021. [Online]. Available: https://doi.org/10.1109/MWC.2021.9690484

[27] Y. Siriwardhana, P. Porambage, M. Liyanage, and M. Ylianttila, "AI and 6G security: Opportunities and challenges," in Proc. Eur. Conf. Netw. Commun. (EuCNC) & 6G Summit, 2021, pp. 1–6. [Online]. Available: https://doi.org/10.1109/EuCNC/6GSummit51104.2021.9482503

[28] P. Porambage et al., "The roadmap to 6G security and privacy," IEEE Open J. Commun. Soc., vol. 2, pp. 663–678, 2021. [Online]. Available: https://doi.org/10.1109/OJCOMS.2021.3078081

[29] H. Du, S. He, L. Su, J. Bai, and R. Yan, "Requirements and potential key technologies of security for 6G mobile network," in Proc. Int. Conf. Netw. Netw. Appl. (NaNA), Qingdao, China, 2023, pp. 1–6. [Online]. Available: https://doi.org/10.1109/NaNA60121.2023.00008

[30] I. Ahmad et al., "Security for 5G and beyond," IEEE Commun. Surv. Tutor., vol. 21, pp. 3091–3133, 2019. [Online]. Available: https://doi.org/10.1109/COMST.2019.2916180

[31] J. Ni, X. Lin, and X. S. Shen, "Efficient and secure service-oriented authentication supporting network slicing for 5G-enabled IoT," IEEE J. Sel. Areas Commun., vol. 36, pp. 598–608, 2018. [Online]. Available: https://doi.org/10.1109/JSAC.2018.2815418

[32] S. Mohammady and D. Malone, "A physical layer security (PLS) approach through address-fed mapping crest factor reduction applicable for 5G/6G signals," in Proc. IEEE Int. Conf. Cloud Comput. Technol. Sci. (CloudCom), Austin, TX, USA, 2022, pp. 1–8. [Online]. Available: https://doi.org/10.1109/FiCloud57274.2022.00047

[33] S. J. Nawaz et al., "Quantum machine learning for 6G communication networks: State-of-the-art and vision for the future," IEEE Access, vol. 7, pp. 29095–29111, 2019. [Online]. Available: https://doi.org/10.1109/ACCESS.2019.2909490

[34] V. Sharma and S. Banerjee, "Analysis of quantum key distribution-based satellite communication," in Proc. IEEE Int. Conf. Commun., Control, Comput. Technol. Smart Grids (ICCCNT), Bangalore, India, 2018, pp. 1–6. [Online]. Available: https://doi.org/10.1109/ICCCNT.2018.8494189

[35] Y. Sun et al., "Blockchain-enabled wireless Internet of Things: Performance analysis and optimal communication node deployment," IEEE Internet

Things J., vol. 6, pp. 5676–5686, 2019. [Online]. Available: https://doi.org/10.1109/JIOT.2019.2905743

[36] F. Tang, Y. Kawamoto, N. Kato, and J. Liu, "Future intelligent and secure vehicular network toward 6G: Machine-learning approaches," Proc. IEEE, vol. 108, pp. 228–245, 2020. [Online]. Available: https://doi.org/10.1109/JPROC.2019.2954595

[37] D. C. Nguyen, P. N. Pathirana, M. Ding, and A. Seneviratne, "Blockchain for 5G and beyond networks: A state-of-the-art survey," J. Netw. Comput. Appl., vol. 166, p. 102693, 2020. [Online]. Available: https://doi.org/10.1016/j.jnca.2020.102693

[38] Z. Yang et al., "Federated learning for 6G: Applications, challenges, and opportunities," Engineering, vol. 8, pp. 1–16, 2022. [Online]. Available: https://doi.org/10.1016/j.eng.2021.12.002

[39] Statista, "Connection density of 4G, 5G, and 6G mobile broadband technologies," 2024. [Online]. Available: https://www.statista.com/statistics/1183690/mobile-broadband-connection-density/. [Accessed: Mar. 11, 2024]

[40] S. A. Ali, R. Hussain, P. Sethi, A. Ahmad, and Y. Kim, "Leveraging machine learning for real-time anomaly detection and self-repair in IoT devices," in Proc. 2023 Int. Conf. Comput. Sci. Artif. Intell. (ICCSAI), Greater Noida, India, 2023, pp. 982–986. doi: 10.1109/ICCSAI59793.2023.10421539

[41] S. Zhu, C. Han, Y. Meng, J. Xu, and T. An, "Embryonics-based phased array antenna structure with self-repair ability," IEEE Access, vol. 8, pp. 209660–209673, 2020. doi: 10.1109/ACCESS.2020.3038857

[42] A. Thompson, P. Layzell, and R. S. Zebulum, "Explorations in design space: Unconventional electronics design through artificial evolution," IEEE Trans. Evol. Comput., vol. 3, pp. 175–196, 1999. doi: 10.1109/4235.788489

[43] A. J. Ferrer, S. Becker, F. Schmidt, L. Thamsen, and O. Kao, "Towards a cognitive compute continuum: An architecture for ad-hoc self-managed swarms," in Proc. 2021 IEEE/ACM 11th Int. Conf. Cloud Comput. Technol. Sci. (CloudCom), 2021, pp. 1–8. doi: 10.1109/CCGrid51090.2021.00076

[44] J. D. K. Hezekiah, A. Singh, M. Ahmad, S. Kumar, and P. Sharma, "Review of next-generation wireless devices with self-energy harvesting for sustainability improvement," Energies, vol. 16, p. 35174, 2023. doi: 10.3390/en16135174

[45] S. M. Rabeek and M. K. Raja, "Design of an autonomous IoT wireless sensor node for industrial environments," in Proc. 2020 Asia-Pacific Microw. Conf. (APMC), 2020, pp. 1–4. doi: 10.1109/APMC47863.2020.9331650

[46] M. Abdel-Basset, G. Manogaran, A. Gamal, and V. Chang, "A novel intelligent medical decision support model based on soft computing and

IoT," IEEE Internet Things J., vol. 7, pp. 5349–5359, 2020. doi: 10.1109/JIOT.2019.2931647

[47] G. Singh, A. Dey, X. Wu, Y. Ma, and Z. Xu, "Near-memory computing: Past, present, and future," Microprocess. Microsyst., vol. 71, p. 102868, 2019. doi: 10.1016/j.micpro.2019.102868

[48] E. Cambria, "Affective computing and sentiment analysis," IEEE Intell. Syst., vol. 31, pp. 102–107, 2016. doi: 10.1109/MIS.2016.31

[49] H. Ma and S. Yarosh, "A review of affective computing research based on function-component-representation framework," IEEE Trans. Affect. Comput., vol. 14, pp. 545–559, 2023. doi: 10.1109/TAFFC.2021.3104512

[50] S. Dustdar, "Distributed computing continuum systems," in Proc. 2022 IEEE Int. Conf. Services Comput. (SCC), Barcelona, Spain, 2022, pp. 356–365. doi: 10.1109/SCC55611.2022.00060

[51] D. Balouek-Thomert, E. G. Renart, A. R. Zamani, A. Simonet, and M. Parashar, "Towards a computing continuum: Enabling edge-to-cloud integration for data-driven workflows," Int. J. High Perform. Comput. Appl., vol. 33, pp. 1159–1174, 2019. doi: 10.1177/1094342019877383

[52] S. Dustdar, V. C. Pujol, and P. K. Donta, "On distributed computing continuum systems," IEEE Trans. Knowl. Data Eng., vol. 35, pp. 1836–1851, 2023. doi: 10.1109/TKDE.2022.3142856

[53] G. Parisi, "Complex systems: A physicist's viewpoint," Physica A: Statistical Mechanics and its Applications, vol. 263, pp. 1–4, 1999. doi: 10.1016/S0378-4371(98)00524-X

[54] K. Friston, J. Kilner, and L. Harrison, "A free energy principle for the brain," Journal of Physiology-Paris, vol. 100, pp. 70–87, 2006. doi: 10.1016/j.jphysparis.2006.10.001

[55] D. Levin and Y. Peres, Markov Chains and Mixing Times, 2nd ed. Providence, RI: American Mathematical Society, 2017

[56] K. Trichias, A. Kaloxylos and C. Willcock, "6G Global Landscape: A Comparative Analysis of 6G Targets and Technological Trends," 2024 Joint European Conference on Networks and Communications & 6G Summit (EuCNC/6G Summit), Antwerp, Belgium, 2024, pp. 1-6, doi: 10.1109/EuCNC/6GSummit60053.2024.10597064

[57] Networld Europe Strategic Research and Innovation Agenda 2022. [online] Available: https://bscw.5g-ppp.eu/pub/bscw.cgi/d516614/SRIA%202022%20Technical%20Annex%20Published.pdf

[58] "Mobile Communications Beyond 2020: The Evolution of 5G Towards the Next G", December 2020, [online] Available: https://www.5gamericas.org/wp-content/uploads/2020/12/Future-Networks-2020-InDesign-PDF.pdf

[59] "6G Applications and Use Cases", [online] Available: https://nextgalliance.org/white_papers/6g-applications-and-use-

cases/#:~:text=Complementing%20the%20audacious%20goals%20the,Communications%2C%20and%20Personalized%20User%20Experiences
[60] "6G The Next Horizon: From Connected People and Things to Connected Intelligence", 2021, [online] Available: https://www.huawei.com/en/huaweitech/future-technologies/6g-the-next-horizon
[61] "Beyond 5G White Paper: Message to the 2030s", March 2022, [online] Available: https://b5g.jp/doc/whitepaper_en_1-0.pdf
[62] "6G Use Cases and Enabling Technologies", [online] Available: https://tsdsi.in/wp-content/uploads/2022/10/6G-White-Paper-12-Pages-Digital.pdf
[63] "TSDSI TR 6017 V1.0.0 6G: Use cases Requirements and Enabling Technologies", [online] Available: https://members.tsdsi.in/s/AGsLX77GtkMx9Y4?dir=undefined&openfile=32663
[64] "MediaTek "6G Vision White Paper"", January 2022, [online] Available: https://d86o2zu8ugzlg.cloudfront.net/mediatek-craft/documents/MediaTek-6G-Vision-White-Paper-EN0122.pdf
[65] S. Alraih, I. Shayea, M. Behjati, R. Nordin, N.F. Abdullah, A. Abu-Samah, et al., "Revolution or Evolution? Technical Requirements and Considerations towards 6G Mobile Communications", Sensors, vol. 22, pp. 762, 2022, [online] Available: https://doi.org/10.3390/s22030762
[66] "Framework and overall objectives of the future development of IMT for 2030 and beyond", November 2023, [online] Available: https://www.itu.int/dms_pubrec/itu-r/rec/m/R-REC-M.2160-0-202311-I%21%21PDF-E.pdf

CHAPTER

2

Micro and Nanotechnologies as Key Enabling Technologies of a Future Network Edge

Jacopo Iannacci

Fondazione Bruno Kessler – FBK, Via Sommarive, 18. Trento – 38123, Italy

Email: iannacci@fbk.eu

Abstract

This chapter develops a comprehensive discussion on the characteristics and features that low-complexity HW components, i.e., SATs, will need to be implemented in order to enable the network edge infrastructure of 6G and FN. To this end, micro and nanotechnologies (MEMS/NEMS) are identified as KETs, capitalizing on their typically high miniaturization, integration and functional diversification, the latter enabled by substantial inexpensiveness of HW redundancy. In particular, a set of 10 founding technology and functional trends are identified. Such items will drive the architecture and evolution of the 6G/FN network edge, and all of them are notably met by micro/nanotechnologies. In brief, the set of 10 technology trends is related to: (1) resilient self-adaptation of the operation and functioning of SATs; broadening and diversification of (2) sensing, (3) actuating, and (4) transducing functionalities; (5) increase in miniaturization and integration of basic physical SAT components; (6) provision, storage and distribution of energy to enable autonomous operation of edge physical nodes; (7) evolution of functions to

DOI: 10.1201/9788743809197-2

cope with altered and unexpected operation conditions; (8) transmission and reception of data, keeping network edge physical nodes linked to the core and orchestrated along with the rest of the infrastructure; and, finally, (9) data storage, and (10) computational capacities, both resident within edge physical nodes, increasing the locally implemented functions and their autonomy with respect to the whole network.

2.1 Introduction

The reference scenario of 6G and FN was addressed through the previous chapter, consolidating the common ground to take the discussion forward. That said, the current chapter moves the focus onto the edge of the network and highlights a comprehensive set of emerging needs related to HW components. Maintaining the attention oriented to low-complexity physical items, e.g., SAT, it is discussed that microtechnologies and nanotechnologies (MEMS/NEMS) are excellent candidates to turn into the KETs of 6G and FN edge smart and autonomous physical devices. The mentioned set of future needs and requested functional features is summarized as follows:

(1) Resilient self-adapting operation, which means the low-complexity HW SAT capacity of self-reacting and self-adapting to varied (and unexpected) operating condition, without the need of a complex system.

(2) Sensing functionalities, which address the need of extending the range of sensed physical magnitudes that can be performed by a unique HW component, as well as by a redundant device realized according to a monolithic approach.

(3) Actuating functionalities, which, similar to the previous item, deals with extending the abilities of a physical actuator, e.g., broadening them, as well as diversifying the implemented actuation principles.

(4) Transduction functionalities, continuing from the previous two points, that is targeting the transducing capacities of basic HW components and mixing sensing and actuating features.

(5) Miniaturization and integration are two technology key drivers bearing the capacity to enable functional diversification of basic physical SAT components.

(6) Energy availability, provision and storage, which start handling the pivotal aspects of powering edge devices and reaching self-sufficiency and node autonomy in term of procurement, storage and availability of energy supporting operations.

(7) Evolution of functions, which, capitalizing on the diversification and integration discussed above, deals with the capacity of basic HW components to implement diverse and orthogonal functions (e.g., sensing and actuation) with respect to the nominal one(s).

(8) Data transmission/reception (Tx/Rx) capacity, which need to be implemented by edge physical nodes and devices, due to the significantly increased demands in terms of data/information transmission from the edge to the core and vice versa.

(9) Data storage, which is inherent to the increasing needs in terms of intelligence distributed and resident within edge network devices.

(10) Computational capacity, which is in full synergy with the previous point and targets the proliferation of features in terms of computation and elaboration performed by 6G/FN network edge devices.

The following pages are going to discuss in more detail each of the ten points listed above.

2.2 Micro/Nanotechnologies as Key Enabling Technologies of the Future Network Edge

After pinpointing the general requirements and possible constraints to the evolution of the future network edge, it is straightforward that reformulation of the classical approaches to the design and development of its physical nodes will be necessary to a certain extent. The resulting landscape of KETs, both SW and HW, turns out to be wide and diversified.

That said, HW technologies for low-complexity physical components, like SATs, are at stake here. Micro and nanotechnologies (MEMS/NEMS), intended as highly miniaturized devices, systems, and innovative micro-/nano-materials and electronics [1]–[4], are identified as suitable candidates for meeting (to a large extent) the demands and constraints of 6G/FN/Super-IoT/TI physical edge nodes.

The way the potential borne by such technologies is reported here is rather practical. To this end, one scheme discussed in [5] is rearranged and reported in Figure 2.1, showing the most relevant demands of 6G/FN network edge nodes, directly and indirectly related to HW technologies.

Each label in Figure 2.1 can be related to partial or total implementation in physical (HW) miniaturized items manufactured in MEMS/NEMS technologies.

This statement will be elaborated in more detail below, including examples of research works in the literature when available. In other cases, consolidated research on some specific themes does not yet exist. However, related contributions are sought and the below is intended as preparatory knowledge for further development of the related item(s) listed above.

Figure 2.1 Most relevant (direct/indirect) HW-related demands of the 6G/FN network edge [5]. (Background image generated by Microsoft Copilot.).

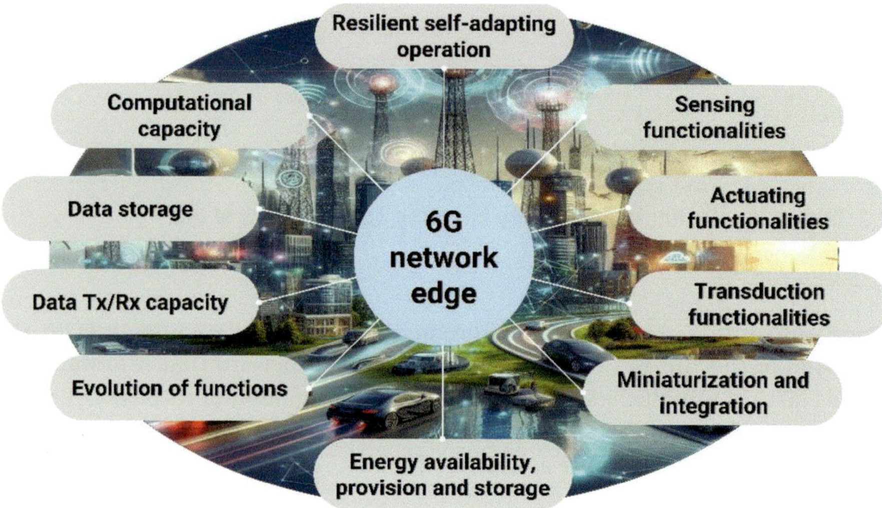

2.2.1 Resilient self-adapting operation

The 6G network edge must adjust its operations in real-time to meet local needs and overcome temporary disruptions. This presents significant challenges for HW performance. E.g., low-complexity HW components must independently respond to changing conditions without relying on dedicated HW–SW systems, which are impractical due to complexity, cost and size constraints. A significant example is in RF channel switching, where current solutions use redundant relays controlled via SW. In contrast, MEMS/NEMS solutions could utilize micro-relays that adapt to RF power levels without the need for dedicated control systems, leveraging the physical properties of thin-films, like, e.g., thermal expansion and self-actuation [6].

To this end, a meaningful example is that of residual stress, along with stress gradients, building within the thin-film during manufacturing for a variety of reasons, as reported in [7]. Residual stress is typically an unwanted effect, since, depending on whether it is predominantly compressive or tensile, can lead to a significant increase in the mechanical stiffness, buckling, as well as non-planarity of suspended membranes. To this end, the microphotograph in Figure 2.2 shows the effects of a stress gradient on an array of suspended cantilevers of various length.

Figure 2.2 Non-planarity of suspended cantilevers manufactured as in [8].

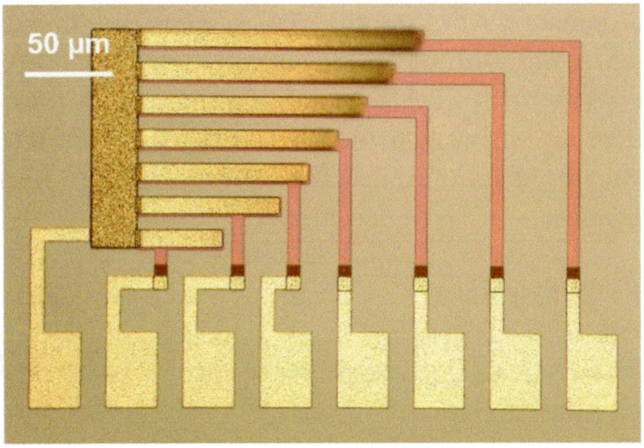

The gradient causes upward bending of the cantilever tips that is increasingly severe as the suspended membrane is longer. For the longer ones, the out-of-plane displacement is so large that the corresponding area is out of focus and appears darker due to the scattering of light away from the microscope illuminator. The test structures in Figure 2.2 are fabricated using the same technology exploited in [8], which capitalizes on residual stress to yield self-assembling out-of-plane structures. This is a relevant example of how an unwanted effect can be leveraged and wisely incorporated in the device functioning.

Along such a direction, the stress release that typically occurs when the membrane is heated up could be exploited to realize micro-devices that adjust their operation on the basis of the increasing heat generated by RF signals carrying more power.

2.2.2 Sensing functionalities

In response to the previous point, sensing functionalities must be expanded. This involves physical sensors performing various and complementary functions, possibly through redundant duplication of different sensors on the same chip or by utilizing a single HW component in multiple ways. These capabilities can be facilitated by microtechnologies and nanotechnologies, leveraging their substantial inexpensiveness, as discussed in [9],[10].

A significant example of what can be achieved by leveraging MEMS and NEMS technologies, is reported in [11]. The discussed MEMS-based chip integrates five different types of sensors in a unique piece of silicon, as they are electrical field sensors, along with accelerometers, pressure, humidity and temperature sensors. Notably, the five sensors are manufactured with a monolithic approach, i.e., with the same micro-fabrication process, and the chip area occupation is 10 × 10 mm^2. Also worth highlighting is that the above listed sensors have different working principles, diverse transduction mechanisms and require radically distant operation conditions. All these features are summarized as follows:

- **The electric field sensor** is a suspended comb-like structure that exploits capacitive coupling for transduction and requires a vacuum to operate.
- **The accelerometer** is a suspended comb-like structure that exploits piezoresistive effect for transduction and requires a vacuum to operate.
- **The humidity sensor** is a device that is not suspended, exploits capacitive coupling for transduction and must be exposed to the external environment.
- **The temperature sensor** is a suspended device, exploits resistive variations for transduction and must be exposed to the external environment.
- **The pressure sensor** is a suspended membrane that exploits the piezoresistive effect for transduction and must be exposed to the external environment.

The way all these orthogonal requirements are addressed is through a complex, yet standard, mixed surface and bulk micromachining MEMS process, which also features a WLP approach with TSV, ensuring hermetic encapsulation for those sensors requiring to operate in a vacuum.

2.2.3 Actuating functionalities

Building upon the prior point, actuators must also expand their functions and modes of operation, similar to sensors. Moreover, integrating mixed sensing and actuating capabilities within a single physical device aligns with the evolving requirements of the 6G network edge. The current state of the art in this area remains constrained, despite a few works, like [12], that have started along this direction. In brief, the work introduces innovative multiple-DoF actuators that capitalize on piezoelectric and electrostatic actuation mechanisms to achieve both in-plane and out-of-plane displacements, aimed at positioning a suspended optical waveguide for chip-to-chip alignment in photonic integrated circuits. The actuator structures, featuring a suspended platform to support the waveguide, are engineered to integrate AlN as the piezoelectric layer for out-of-plane actuation, and a comb-drive with interdigitated fingers placed on the same layer for in-plane movement. Two separate configurations, i.e., a 2-DoF model with Z-axis and Y-axis motion, and a 3-DoF model with motion along the three axes, were developed and experimentally validated. Both versions incorporate capacitive displacement sensors for monitoring motion along the Z- and Y-axes, i.e., stepping in the direction of mixed sensing and actuating capacities in a unique miniaturized low-complexity HW component.

2.2.4 Transduction functionalities

Transduction between different physical domain is an inherent feature of sensors and actuators. However, devices primarily focused on transduction capabilities, rather than sensing or actuating, are anticipated to be crucial, as well. This includes technologies like RF-MEMS/-NEMS, i.e., radio frequency MEMS/NEMS [13],[14] and EH-MEMS/-NEMS, i.e., MEMS/NEMS for EH [15],[16], which have been extensively explored in the literature. However, there remains a gap in hybridizing these technologies to enable a single hardware unit to perform its function while simultaneously harvesting the required energy for operation. A couple of discussions venturing such direction are elaborated in [17],[18]. A microphotograph of a physical RF-MEMS complex switching device, manufactured in the technology discussed in [19], is shown in Figure 2.3.

Figure 2.3 Microphotograph of a complex RF-MEMS switching device [19].

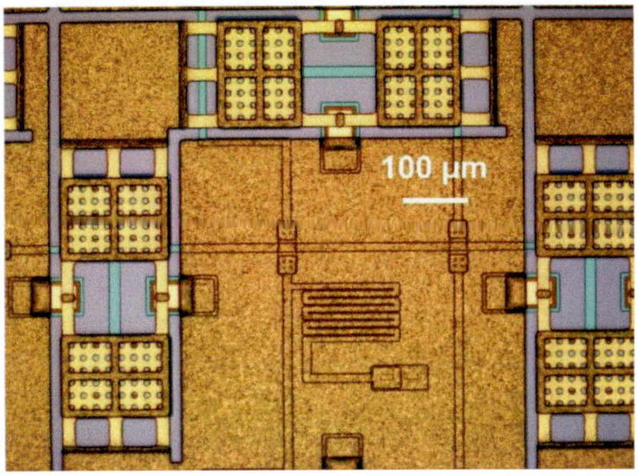

A photo of an entire silicon wafer populated by diverse EH-MEMS design concepts targeting mechanical vibrations, as discussed in [20], is shown in Figure 2.4.

Figure 2.4 Photo of a silicon wafer with diverse EH-MEMS designs [20].

2.2.5 Miniaturization and integration

The conduit linking all the previous points, as well as those in the following, is the relentless need for size reduction of physical devices at the edge, with a parallel increase of functionalities. That being said, micro and nanotechnologies trigger ample opportunities for advancing miniaturization, packaging, integration, and the fusion of functions within a single HW unit [21],[22]. Focusing on [22], the work discusses high-density fan-out WLP technologies, focusing on two key solutions, namely WL-SiP and 3D SWIFT. Such solutions enable efficient integration of diverse components, such as, e.g., logic, memory, power IC, RF and passive elements, into compact, high-performance packages suitable for next-generation applications in various segments, like mobile devices, automotive and HPC.

In more detail, WL-SiP allows for WL integration of multi-functional dies exploiting copper as a redistribution layer. This yields reduced package thickness and improved thermal and electrical performance. On the other hand, 3D SWIFT targets ultra-thin 3D heterogeneous integration by eliminating traditional substrates and stacking dies directly onto the surface with a patterned redistribution layer. This solution enables thinner form factors, shorter interconnect paths, and improved warpage behaviour without using underfill materials. Additionally, [22] explores large-body MCMs using

TSV-less interposers as an alternative to 2.5D silicon interposers, the latter suffering from cost and supply chain limitations. The SWIFT MCM facilitates ASIC and memory integration, supporting various applications, e.g., high-speed networking and server applications.

2.2.6 Energy availability, provision and storage

As previously emphasized, energy management will be pivotal at the 6G edge. Scientific research is advanced in transducer technologies for converting environmental energy (e.g., EH in Section 0), transferring it (e.g., WPT [23]), and storing it (e.g., miniaturized batteries [24]).

However, there is a need to integrate and align these technologies to achieve the concept of EADO [5],[25]. Despite rich contributions in individual technology branches, there remains a gap in inclusive HW platform solutions that encompass diverse energy converters, optimized extraction and storage, and adaptive operation strategies tailored to real-time power availability. An example is discussed in [26]. The work reports a micro power generator that simultaneously harvests thermal and optical energy by integrating a micro-TEG and a solar cell on a single chip. The device is designed for low-power applications such as IoT nodes, and it leverages the advantages of solid-state EH, i.e., absence of moving parts, high reliability, and long operational life, within a compact footprint. The fabrication is fully compatible with standard CMOS technology, allowing low-cost, batch processing and scalable integration. The proposed device incorporates a hybrid structure where a thermopile and a solar cell are layered to optimize space and energy collection efficiency. To improve TEG performance, the device uses a polyimide insulating layer and oxide passivation to enhance thermal resistance and direct heat flow. The thermopile comprises alternating n-type and p-type polycrystalline silicon legs connected via gold strips, with the entire structure thermally insulated to maximize temperature gradients. The solar cell, located at the top of the chip, features interdigitated polycrystalline silicon electrodes and a selectively perforated oxide layer to reduce carrier recombination and reflection losses.

2.2.7 Evolution of functions

Building on the earlier Section 2.2.1, the self-adaptation of 6G extends beyond mitigating local challenges. With extensive use of AI, services and functionalities will evolve, potentially incorporating unplanned features. This expanded flexibility cannot solely rely on SW, particularly at the network edge, where extensive HW/SW resources are limited and not fully accessible, as widely discussed in Chapter 1.

Bearing in mind these considerations, and capitalizing on Section 0 discussing energy aspects, a crucial further step is that of incorporating EH functionalities within a physical sensor or transducer, therefore moving in the direction of self-powered basic HW components. To this end, a significant example available in the literature is a miniaturized tri-axial inertial sensor, featuring a seismic cylindrical mass and four suspending slender beams. The schematic is reported in Figure 2.5 and its technical details are thoroughly discussed in [17],[27],[28].

Figure 2.5 Schematic of the self-powered tri-axial accelerometer in [17],[27],[28].

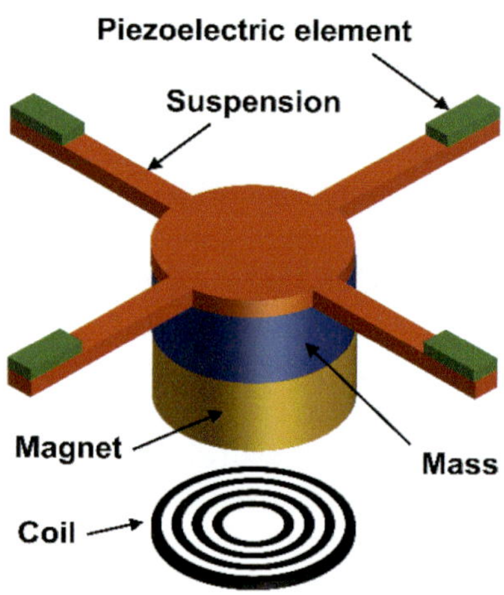

The device has been miniaturized to a certain degree, although it is not produced using micro-fabrication (MEMS) methods. Its size is approximately 110 × 110 × 20 mm^3. Nonetheless, there remains ample potential for further downsizing and integration using MEMS technology. A key technical aspect of the design lies in the simultaneous use of piezoelectric and electromagnetic transduction principles. Specifically, the piezoelectric mechanism enables acceleration sensing by converting mechanical strain into corresponding electric charge. In contrast, the electromagnetic mechanism implements EH from the mass–spring system mechanical movements. The harvested vibration energy powers the charge amplifier that processes the accelerometer output, along with the control logic interfacing with the inertial sensors. The latter logic also compensates the impact of the electromagnetic EH, which inherently alters the sensor dynamic response, resulting in a self-sufficient sensing HW sub-system. Initial simulations show that the energy harvester can generate 10 mW under horizontal vibrations at 100 Hz.

2.2.8 Data transmission/reception (Tx/Rx) capacity

In contrast to earlier generations, 6G will bring a disruptive shift from centralized (cloud) to distributed (fog/edge) capabilities (see Chapter 1). Yet, this transition does not diminish the importance of data transmission and reception across the network periphery and core. Edge HW infrastructure nodes will require high-performance, versatile and ULP RF transceivers (transmitters/receivers). MEMS/NEMS solutions are expected to offer significant advantages in providing the necessary RF passive components for such systems (see Section 0).

From a data transmission and communication standpoint, 6G will build further upon mMIMO antenna technologies, enabling sophisticated beamforming within small-, pico-, and nano-cell environments [29], as schematically reported in Figure 2.6.

Figure 2.6 Schematic of a mMIMO antenna with advanced beamforming [25].

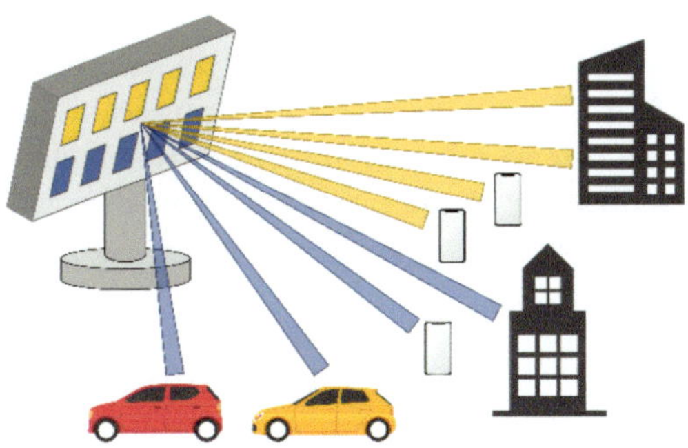

The development of compact, high-order antenna arrays is progressing steadily. However, this represents only a portion of the overall system complexity. As a matter of fact, mMIMO systems require complex and redundant HW to operate effectively in both Tx and Rx modes. As a result, the TRM must be positioned between the mMIMO and the rest of the RF chain and then physically replicated for several antenna elements (ideally for each of them), potentially reaching configurations of 64 × 64 elements or beyond. Furthermore, operation in the sub-THz band (100–300 GHz) significantly complicates the design challenges of high-performance TRM components, especially concerning losses, energy efficiency, fabrication costs, and integration.

Digging deeper into the detail, the TRM sub-unit named PCU integrates tunable elements such as PAt and PhS, which manage the amplitude and delay of the RF signal, respectively. It also includes the SwU selecting the Tx or Rx mode of each individual antenna, as illustrated in the block diagram in Figure 2.7.

Figure 2.7 Diagram of a mMIMO system, with TRM and sub-blocks (PCU/RFFE) [25].

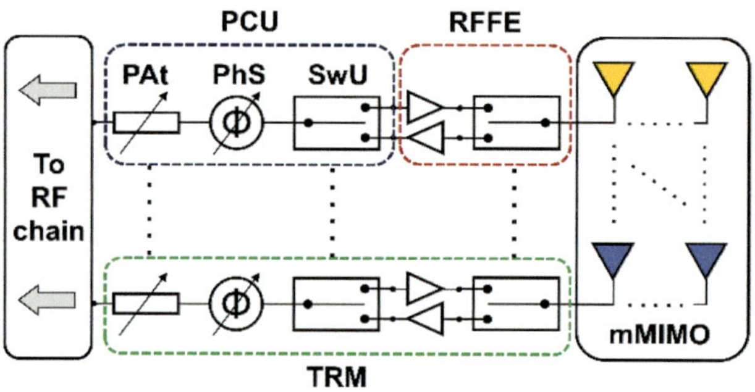

RF-MEMS technology has already proven its suitability in implementing high-performance, broadband and highly reconfigurable PAt, PhS and SwU. Leveraging the strong integration capabilities of this technology, the prospect of fabricating a monolithic PCU on a single silicon chip can become tangible, as noted in [25].

2.2.9 Data storage

Combining the trends of resource distribution, AI utilization, and self-adaptive operational evolution, alongside the importance of local data storage capacities, the need for miniaturized, ultra-low power consumption, and very-low access latency HW components becomes paramount.

Leveraging existing and emerging micro and nanotechnology-based solutions, such as, e.g., [30]–[32], in a synergic way could prove to be crucial in meeting these demands, as also emphasized in Section 2.2.10.

Stepping into a little more detail, MEMS and NEMS technologies have been explored for over a decade to develop highly integrated, non-volatile memory units with ULP consumption. As an example, [33] demonstrates a MEMS mechanical resonator that performs both logic and memory functions. It differentiates between the "0" and "1" logic states using the nonlinear hysteresis typical of resonators, when the excitation frequency is swept across the resonance. A similar concept is applied in [30], in the latter case using NEMS technology, thus enabling further miniaturization.

Another approach involves electrostatically actuated NEMS nano-switches, designed as cantilevers or clamped–clamped membranes. These devices utilize pull-in/pull-out hysteresis, where logic states correspond to actuation and release voltage thresholds. Within this context, a notable advancement appears in [31], where NEMS nano-relays are fabricated using a few CMOS-compatible steps. Traditional electrostatic switches are monostable, requiring continuous DC bias to hold the actuated state. That said, [31] introduces trapping of charge beneath the NEMS membrane, shifting the hysteresis curve toward 0 V and enabling, in turn, bi-stability. This allows the device to maintain each of its states without constant power, requiring energy only during state programming, i.e., state transition.

2.2.10 Computational capacity

In connection with the preceding point and with the discussion in Chapter 1, there will be a significant rise in demands for high-efficiency edge computation capacities. Within this context, the two emerging research streams of CIM and of non-conventional computing will exert significant impact. These research areas, although not fully explored, align well with the capabilities offered by micro/nanotechnologies. Examples of investigations along both such directions are discussed in [33] and [34], respectively.

On a diverse reference plane, once again realizing multiple orthogonal functions through a single low-complexity HW element is still a pivotal aspect. Within such a context, embedding intelligence can be related to the ability of a basic sensor or actuator to carry out (basic) signal conditioning and/or processing of the sensed quantity. In this sense, the selected example is a MEMS-based inertial sensor, as described in [35] and schematically illustrated in Figure 2.8.

The concept introduces a self-computing digital accelerometer, effectively eliminating the need for dedicated readout electronics. This is achieved by cascading inertial MEMS ohmic switches, with the number of micro-relays determining the resolution in bits. Each switch proof mass is connected to a flexible suspension with a distinct spring constant. In the 3-bit simplified model shown in Figure 2.8, the spring constant k is tuned by changing the width of the suspension beam supporting the seismic mass. In particular, it is set to $k/4$ for bit 0 (LSB), $k/2$ for bit 1, and k for bit 2 (MSB). The output corresponding to a specific acceleration level is mapped by individual voltage drops associated with the sequential activation of the switches across the sensor micro-relay array. This way, the inertial sensor autonomously delivers a discretized (digital) voltage output for each measured acceleration level, thus waiving the need for external readout electronics to perform this conditioning step.

Figure 2.8 Schematic of the 3-bit accelerometer in [35], based on coupled switches and binary output, with reference to (a) bit 0 (LSB); (b) bit 1; (c) bit 2 (MSB).

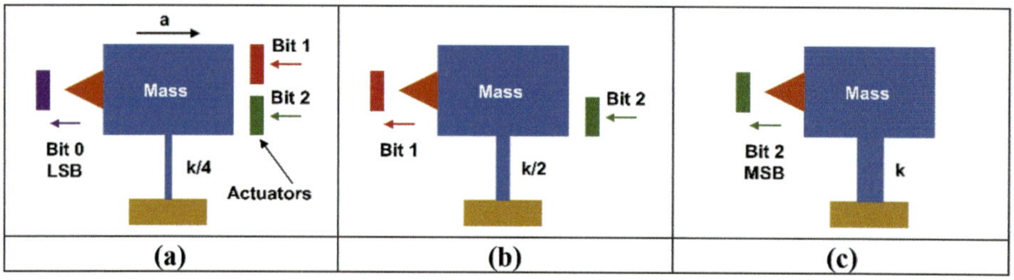

2.3 Conclusion

In conclusion, this chapter details the crucial characteristics and features required for low-complexity HW components, such as SATs, to enable the network edge infrastructure of the 6G and FN. The analysis has marked the significance of micro and nanotechnologies (MEMS/NEMS) as KETs, emphasizing their intrinsic strengths in terms of miniaturization, integration and diversification of functionalities through cost-effective HW redundancy. The analysis led to the identification of ten fundamental technology and functional trends that are expected to drive the development of the 6G/FN network edge. In this sense, the chapter has demonstrated the strong alignment between the capabilities offered by micro/nanotechnologies and the set of identified trends, which include resilient self-adaptation, a wider range of sensing, actuating and transducing functionalities, further increased miniaturization and integration, autonomous energy management, evolution towards adaptive operations, efficient data transmission and reception, and enhanced local data storage and computational resources.

References

[1] M. Gad-el-Hak, Ed., MEMS – Introduction and Fundamentals, 1st ed. Boca Raton, FL, USA: CRC Press, 2005, pp. 1–492

[2] M. Nasrollahzadeh, Z. Issaabadi, M. Sajjadi, S. M. Sajadi, and M. Atarod, "Types of nanostructures," in An Introduction to Green Nanotechnology, vol. 28, 2019, pp. 25–56

[3] B. K. Kaushik, Nanoelectronics: Devices, Circuits and Systems. CRC Press, 2018, pp. 1–300

[4] Y. Gogotsi, Nanomaterials Handbook, 2nd ed. CRC Press, 2017, pp. 1–1200.

[5] J. Iannacci and H. V. Poor, "Review and perspectives of micro/nano technologies as key-enablers of 6G," IEEE Access, vol. 10, pp. 55428–55458, 2022. doi: 10.1109/ACCESS.2022.3176348

[6] L. Ni and M. P. De Boer, "Self-actuating isothermal nanomechanical test platform for tensile creep measurement of freestanding thin films," J. Microelectromech. Syst., vol. 31, pp. 128–136, 2022. doi: 10.1109/JMEMS.2021.3131153

[7] M. Huff, "Review Paper: Residual Stresses in Deposited Thin-Film Material Layers for Micro- and Nano-Systems Manufacturing," Dec. 01, 2022, *MDPI*. doi: 10.3390/mi13122084

[8] A. Bagolini, F. Giacomozzi, B. Margesin, and P. Bellutti, "A single metal layer MEMS self-assembling coplanar structure," J. Micromechanics Microengineering, vol. 17, no. 5, 2007, doi: 10.1088/0960-1317/17/5/014

[9] M. Hautefeuille, B. O'Flynn, F. Peters, and C. O'Mahony, "Miniaturised multi-MEMS sensor development," Microelectron. Reliab., vol. 49, pp. 772–778, 2009. doi: 10.1016/j.microrel.2009.02.017

[10] M. Mansoor et al., "SOI CMOS multi-sensors MEMS chip for aerospace applications," in Proc. IEEE Sensors, 2014, pp. 1132–1135. doi: 10.1109/ICSENS.2014.6985225

[11] Qing Ma, Z. Wang and Liyang Pan, "Monolithic integration of multiple sensors on a single silicon chip," 2016 Symposium on Design, Test, Integration and Packaging of MEMS/MOEMS (DTIP), Budapest, 2016, pp. 1-4, doi: 10.1109/DTIP.2016.7514831

[12] A. A. S. Rabih et al., "Multi degrees-of-freedom hybrid piezoelectric-electrostatic MEMS actuators integrated with displacement sensors," IEEE J. Microelectromech. Syst., vol. 33, pp. 21–36, 2024. doi: 10.1109/JMEMS.2023.3341039

[13] N. Zhang, R. Song, J. Liu, and J. Yang, "A packaged THz shunt RF MEMS switch with low insertion loss," IEEE Sensors J., vol. 21, pp. 23829–23837, 2021. doi: 10.1109/JSEN.2021.3113647

[14] J. Iannacci, "RF-MEMS technology as an enabler of 5G: Low-loss ohmic switch tested up to 110 GHz," Sens. Actuators A Phys., vol. 279, pp. 152 157, 2018. doi: 10.1016/j.sna.2018.07.005

[15] B. Debnath and R. Kumar, "A comparative simulation study of the different variations of PZT piezoelectric material by using a MEMS vibration energy harvester," IEEE Trans. Ind. Appl., vol. 58, pp. 6747–6755, 2022. doi: 10.1109/TIA.2022.3160144

[16] J. Xie, C. Lee, and H. Feng, "Design, fabrication, and characterization of CMOS MEMS-based thermoelectric power generators," IEEE J. Microelectromech. Syst., vol. 19, pp. 317–324, 2010. doi: 10.1109/JMEMS.2010.2041035

[17] Y. Zhu, D. F. Wang, Y. Fu, and X. Yang, "Developing self-powered high-performance sensors: Part I - A tri-axial piezoelectric accelerometer with an error compensation method," in Proc. DTIP, 2018, pp. 1–4. doi: 10.1109/DTIP.2018.8394223

[18] Z. Huo, Y. Wei, Y. Wang, Z. L. Wang, and Q. Sun, "Integrated self-powered sensors based on 2D material devices," Adv. Funct. Mater., vol. 32, p. 2208329, 2022. doi: 10.1002/adfm.202206900

[19] F. Giacomozzi, V. Mulloni, S. Colpo, J. Iannacci, B. Margesin, and A. Faes, "A flexible fabrication process for RF MEMS devices," Rom. J. Inf. Sci. Technol., vol. 14, pp. 285–294, 2011

[20] J. Iannacci, G. Sordo, E. Serra, and U. Schmid, "The MEMS four-leaf clover wideband vibration energy harvesting device: Design concept and experimental verification," Microsyst. Technol., vol. 22, pp. 1775–1784, 2016. doi: 10.1007/s00542-016-2886-3

[21] S. Li, SiP-System in Package Design and Simulation. Wiley, 2017, pp. 1–348. doi: 10.1002/9781119045991

[22] K. Lee, "High-density fan-out technology for advanced SiP and 3D heterogeneous integration," in Proc. IEEE RPS, 2018, pp. 4D.1-1–4D.1-4. doi: 10.1109/IRPS.2018.8353588

[23] N. Xing and G. A. Rincón-Mora, "Highest wireless power: Inductively coupled or RF?," in Proc. ISQED, 2020, pp. 298–301. doi: 10.1109/ISQED48828.2020.9136990

[24] A. Kornyushchenko, V. Natalich, S. Shevchenko, and V. Perekrestov, "Formation of Zn/ZnO and Zn/ZnO/NiO multilayer porous nanosystems for potential application as electrodes in Li-ion batteries," in Proc. IEEE NAP, 2020, pp. 01NSSA06-1–01NSSA06-5. doi: 10.1109/NAP51477.2020.9309614

[25] J. Iannacci, "A perspective vision of micro/nano systems and technologies as enablers of 6G, super-IoT, and tactile internet," Proc. IEEE, vol. 111, pp. 5–18, 2023. doi: 10.1109/JPROC.2022.3223791

[26] J. Yan, X. Liao, S. Ji, and S. Zhang, "A novel multi-source micro power generator for harvesting thermal and optical energy," IEEE Electron Device Lett., vol. 40, pp. 349–352, 2019. doi: 10.1109/LED.2018.2889300

[27] L. Zheng, D. F. Wang, Y. Fu, X. Yang, Y. Suzuki and M. Ryutaro, "Developing Self-powered High Performance Sensors: Part II - Preliminary Study On PE-ME Coupling In A Vibration Energy Convertor," 2020, doi: 10.1109/DTIP51112.2020.9139142

[28] C. Li, Q. Lan, D. F. Wang, T. Itoh, and R. Maeda, "Developing Self-powered High Performance Sensors: Part III-A Low Frequency Ball-impacted PE/ME Composite Energy Harvester," 2022, doi: 10.1109/DTIP56576.2022.9911741

[29] W. Jiang and F.-L. Luo, 6G Key Technologies: A Comprehensive Guide, 2023, doi: 10.1002/9781119847502

[30] A. Uranga et al., "Exploitation of non-linearities in CMOS-NEMS electrostatic resonators for mechanical memories," Sens. Actuators A Phys., vol. 197, pp. 88–95, 2013. doi: 10.1016/j.sna.2013.03.032

[31] W. W. Jang et al., "NEMS switch with 30 nm-thick beam and 20 nm-thick air-gap for high-density non-volatile memory applications," Solid State Electron., vol. 52, pp. 1578–1583, 2008. doi: 10.1016/j.sse.2008.06.026

[32] K. Choe, J. Park, and C. Shin, "Theoretical study of ferroelectric-gated nanoelectromechanical diode nonvolatile memory cell," Solid State Electron., vol. 163, p. 107662, 2020. doi: 10.1016/j.sse.2019.107662

[33] A. Yao and T. Hikihara, "Reprogrammable logic-memory device of a mechanical resonator," Int. J. Nonlinear Mech., vol. 94, pp. 406–416, 2017. doi: 10.1016/j.ijnonlinmec.2016.11.011

[34] G. Martini, M. Mirigliano, B. Paroli, and P. Milani, "The receptron: A device for the implementation of information processing systems based on complex nanostructured systems," Jpn. J. Appl. Phys., vol. 61, no. SM, p. 040305, 2022. doi: 10.35848/1347-4065/ac665c

[35] V. Kumar, X. Guo, R. Jafari, and S. Pourkamali, "Ultra-low power self-computing binary output digital MEMS accelerometer," in Proceedings of the IEEE International Conference on Micro Electro Mechanical Systems (MEMS), 2016, vol. 2016-February, doi: 10.1109/MEMSYS.2016.7421607

CHAPTER

3

Micro/Nano Fabrication Processes: Utilizing Borrowed and Dedicated Fabrication Processes for 6G and Future Network Devices

Praful P. Kabadi[1], Rajesh K. Singh[1], Bhubon Chandra Mech[1] and Vinod Belwanshi[2,3,4,*]

[1]Defence Institute of Advanced Technology (DIAT), Pune, India 411025
[2]CSIR-National Metallurgical Laboratory, Jamshedpur, India 831007
[3]Academy of Scientific and Innovative Research, Ghaziabad, India 201002
[4] School of Physics and Astronomy, University of Glasgow, UK G12 8QQ

Email: *vinodbelwanshi.nml@csir.res.in

Abstract

Micro/nanofabrication techniques and processes play a vital role that is crucial in enabling miniaturised and lightweight compact devices and systems. The ability to have precise and reliable fabrication with sub-10 nm features is the one of the key requirements for driving innovation and advancement in nanoscience and nanotechnology, and it is also indispensable for many non-CMOS devices. While conventional lithography is a mainstream approach, it is limited in resolution, and precision for realising the micro/nano devices that are being targeted for use in the 6G future network. Unconventional methods like mechanics-enabled and post-trimming methods provide complementary advantages by providing control in microstructures geometry, material

DOI: 10.1201/9788743809197-3

adaptability, and precision tuning beyond the capabilities of conventional lithographic methods. These unconventional methods are suitable for low temperature applications. The post-trimming techniques includes focused ion beam (FIB) milling, atomic layer etching, thermal or chemical trimming and electrochemical tuning. The chapter explores fabrication methods for advanced micro/nano devices [3], including synthesis, on-chip transfer, film patterning, and property tuning. It also explores the fabrication methodology of dielectric metasurfaces using standard nanolithography techniques and extreme nanofabrication, highlighting methods like EBL, FIB, laser-based nanomanufacturing, greyscale lithography, multistep lithography, nanoimprint lithography, deep reactive ion etching and many other techniques that are borrowed from integrated circuit (IC) technology.The application of 3D printing for electromagnetic metamaterial absorbers is also discussed. The context of these fabrication innovations includes potential applications in 6G wireless networks, particularly terahertz (THz) communications, where reconfigurable intelligent surfaces (RISs) and metasurfaces are proposed to address challenges like signal loss. Fabrication is also critical for emerging applications such as quantum devices, nano-optics, and high-performance sensing, often requiring sub-10 nm features and novel materials. The chapter will also touch upon MEMS-based devices and their integration, and the challenges associated with these advanced fabrication techniques, such as scalability, throughput, and their cost effectiveness.

3.1 Introduction

Micro/nanofabrication techniques have gained attention to realise miniaturised and low weight tiny devices. These techniques are essential key enablers for creating miniaturised functionalities used in various applications, including energy harvesting/storage, bio/chemical sensing, solar cells, and lighting sources [1]. Advancements in micro/nanofabrication are observed to be as an essential driving force for many technologies, such as integrated circuits (ICs), energy, medicine, health monitoring, and information storage. Reliable fabrication of micro/nanostructures with sub-10 nm features is of great significance for advancing nanoscience and nanotechnology for the current demand of 6G devices and future networks [2]. Although structures on the sub-10 nm scale can be produced by current complementary metal-oxide semiconductor (CMOS) chip manufacturing, many emerging applications need ultrasmall features where CMOS methods are not always practical or appropriate due to factors like exotic features, material compatibility, and usage even cost. As a result, numerous specialized techniques have been created for varied uses [2].

Reliable manufacturing at the sub-10 nm scale is far more difficult than for feature sizes bigger than 10 nm, where fabrication techniques are well developed. With features like nanofins in FinFET technology reducing to 7 nm, the IC industry's node has shrunk to the single-digit nanometre scale. However, because of the high cost and problems with process compatibility, this process portfolio cannot be immediately translated to research and development of innovative devices. Because of their innovative materials and unusual layouts, these emerging devices require many unconventional approaches. Both the IC industry and the research and development of micro/nanoelectronic, nano-optical, biomedical, and quantum devices have expressed a strong desire to continuously advance nanofabrication capabilities to the sub-10 nm scale. Additionally, nanofabrication offers a platform for the development of synthetic materials and metadevices for the small-scale investigation of novel physical phenomena [2].

The micro/nano devices for 5G/6G communication applications demand advanced microfabrication technology along with conventional techniques. These devices are broadly classified into the following categories. The microdevices that operate at mmWave or THz frequencies and form the core of wireless transceivers. This includes RF MEMS switches, micromachined phase shifters, RF power amplifiers (CMOS/GaN-based), microwave filters (SAW/BAW/MEMS), low noise amplifiers (LNAs), micromachined transmission lines (e.g., CPW, microstrip), and micromachined waveguides for mmWave/THz. Antena and beamforming microdevices enable advanced antenna performance and dynamic beam steering which includes MEMS-based reconfigurable antennas, on-chip mmWave/THz antennas, MEMS beam steering units, metamaterial-based micro antennas, planar array antenna chips and reconfigurable intelligent surface (RIS) tiles. The 6G/5G networks also demand processing signals at the physical layer and interface with digital systems. The mixers and modulators (CMOS/SOI), voltage-controlled oscillators (VCOs), phase-locked loops (PLLs), RF frequency synthesizers, tunable matching networks (using MEMS/varactors), and front-end ICs (FEICs) for mmWave/THz are a few microdevices used in signal processing networks. These networks also demand intelligent sensing and environmental parameter sensors which include microbolometers for THz detection, MEMS environmental sensors such as pressure, humidity, gas for edge-aware networks, MEMS accelerometers and gyroscopes for positioning in 6G, THz detectors and emitters, and optoelectronic microdetectors. Figure 3.1 shows an overview of micro/nano devices for 6G.

Figure 3.1 Overview of micro/nano devices for 6G.

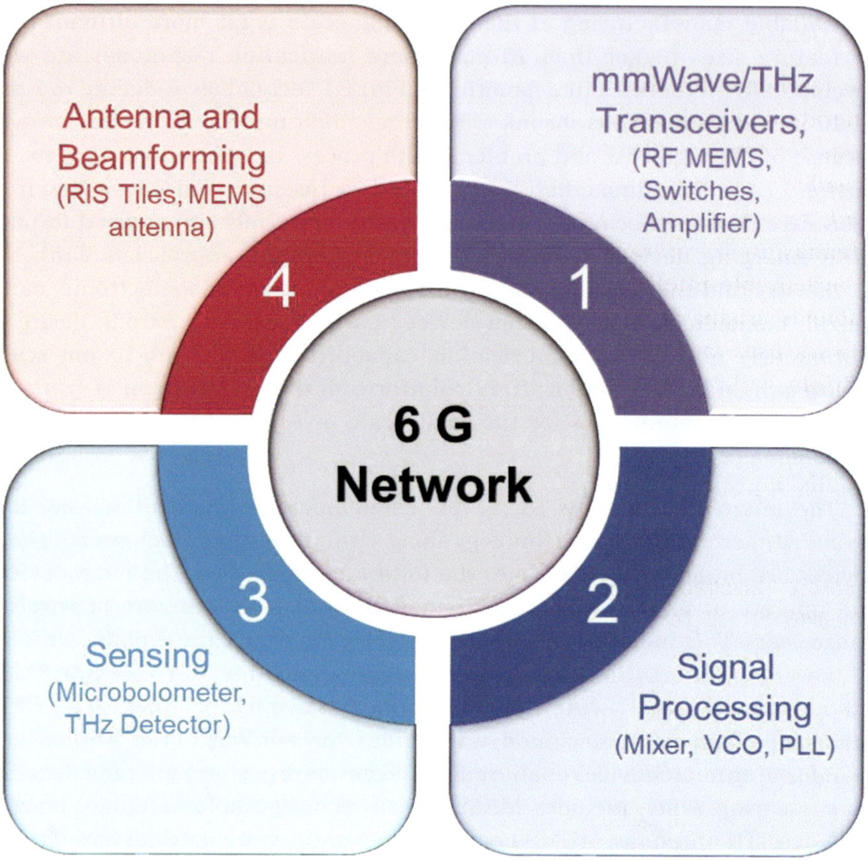

The purpose of this chapter is to present fabrication techniques that are pertinent to advanced micro/nano devices, such as methods for film patterning, on-chip transfer, material synthesis, and property tuning/modification for materials such as 2D materials. We will discuss how to fabricate metasurfaces and metamaterial absorbers as well as how to implement 2D van der Waals heterostructures on chips [4]. Both conventional and non-traditional methods will be discussed, including those taken from established sectors like CMOS and MEMS as well as specialized procedures like 3D printing and particular lithography techniques.

3.2 The Need for Fabrication Innovation for 6G

3.2.1 Terahertz communication demands

The terahertz (0.1–10 THz) spectrum is anticipated to be used by the next generation of wireless networks (6G and beyond) to satisfy future application needs for high bandwidth and low latency. Molecular absorption and large spreading and penetration losses make it difficult to construct systems in this band, requiring strong line-of-sight requirements and extremely directional antennas. By rerouting incident waves and establishing a virtual line-of-sight, reconfigurable intelligent surfaces (RISs) have been offered as a solution to problems arising from non-line-of-sight propagation [6], [7]. However, if the network uses more than one band, the advantages of RISs can be lessened. Multiple wideband transparency windows are required for RISs, or more generally, metasurfaces [6]. An illustration of terahertz communication challenges and the role of reconfigurable intelligent surfaces (RISs) in establishing a virtual line-of-sight is depicted in Figure 3.2.

Figure 3.2 Illustration of terahertz communication challenges and the role of reconfigurable intelligent surfaces (RISs) in establishing a virtual line-of-sight.

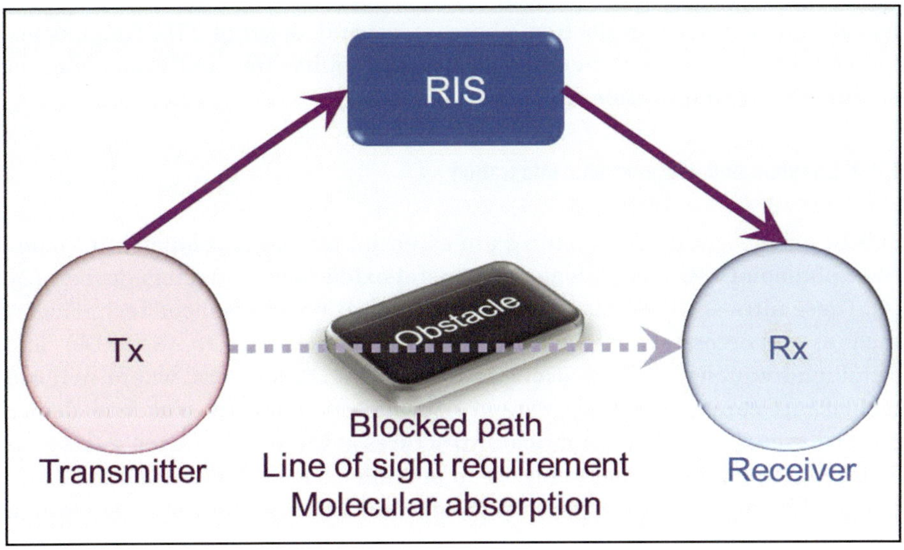

3.2.2 Terahertz antenna arrays and beam formers

Terahertz (THz) antennas and beamforming technologies are recognized as critical components for the advancement of end-to-end THz communication (THz Com) systems. Among these, graphene-based antennas have garnered particular attention due to their unique electromagnetic properties and tunability at THz frequencies [9]. In parallel, metasurface (MS) engineered two-dimensional structures composed of subwavelength unit cells play a vital role in manipulating electromagnetic waves. The overall electromagnetic response of a metasurface, especially a reflective metasurface, is governed by the collective behaviour of its unit cells. For optimal performance, these unit cells must deliver high reflection amplitude and offer precise control over the phase of reflected waves across a broad phase range [6].

Achieving dynamic reconfigurability, the ability to adapt the response of the metasurface in real time requires meeting these performance criteria without altering the physical geometry of the structure [6]. This can be accomplished in the THz regime by exploiting the electrostatic tunability of graphene. Specifically, adjusting the biasing voltage applied to graphene sheets integrated within each unit cell enables active control over both the operating frequency and the phase response. Furthermore, incorporating multiple graphene layers within the unit cell design offers even greater versatility. By independently tuning the biasing voltage of each layer, it becomes possible to dynamically reconfigure the local reflection phase at several discrete THz frequencies, thereby enhancing the metasurface's adaptability for multifunctional or broadband THz applications [6].

3.2.3 Quantum and neuromorphic integration

Sub-10 nanometre (nm) structures are essential for the development of many emerging non-CMOS (complementary metal-oxide-semiconductor) devices [2], [4]. These ultra-small features enable the realization of advanced technologies such as superconductor nanowire single-photon detectors (SNSPDs) and ultrahigh-frequency surface acoustic wave (SAW) resonators, where reduced dimensions can significantly extend operational ranges or enhance device performance. The ability to fabricate structures at the sub-10 nm scale serves as a key enabling platform, not only for improving existing device functionalities but also for exploring novel physical phenomena and designing next-generation concept devices [2]. Such fine-scale fabrication techniques open diverse application areas including enhanced spectroscopy, high-sensitivity sensing, field emission technologies, genetic sequencing, and various quantum devices.

Notably, devices such as the single-electron transistor (SET) and radio-frequency single-electron transistor (RF-SET) are fundamentally dependent on quantum tunnelling across nanometre-scale gaps, demonstrating the critical role of ultra-small features in enabling these effects. Moreover, quantum confinement a phenomenon where charge carriers are restricted to very small dimensions underpins applications like integrated quantum light sources using nitrogen-vacancy (NV) centres in photonic crystal cavities, graphene nanoribbon field-effect transistors (FETs), and silicon quantum dots [2].

In all these cases, the fabrication of precise sub-10 nm nanogaps and nanoscale features is not just advantageous but fundamental to the realization and performance of quantum-based and nanoscale devices [2].

3.3 Borrowed Fabrication Processes

3.3.1 CMOS-compatible processes

The capability of current complementary metal-oxide semiconductor (CMOS) chip manufacturing can produce structures on the sub-10 nm scale [2]. The node in the latest silicon (Si)-based ICs manufacturing industry has achieved the sub-10 nm scale by combining sophisticated lithography, etch, and film deposition processes [2]. The architecture of field-effect transistors (FETs) has evolved from planar to FinFETs, with nanofin widths shrinking to 7 nm [2]. CMOS-compatible fabrication technology of silicon allows for efficient production of nanostructures, both horizontally and vertically [4]. Figure 3.4 in [4] illustrates smooth sidewalls of silicon rectangle pillars fabricated after reactive ion etching (RIE) [4]. However, the process portfolio in the Si-based IC industry cannot be directly transferred to applications related to the research and development of novel devices due to the extremely high cost and process compatibility [2]. CMOS-based manufacturing methods are currently not feasible or appropriate for many emerging applications requiring sub-10 nm features, novel materials, or irregular layouts [2].

In general, CMOS technology is foundational for modern semiconductor devices, enabling the integration of both n type and p-type transistors on a single chip. The fabrication process involves several key steps starting from substrate preparation, typically, a p-type silicon wafer is used as a substrate material, which needs to be cleaned using a set of RCA cleaning steps. A thin layer of high-quality silicon (epitaxial process) is grown on the wafer to ensure uniformity and control over doping concentrations. The cleaned silicon wafers undergo thorough thermal oxidation either by wet or dry methods, depending on the requirement. The oxide layer act as a heat of the micro/nano devices which provides a good isolation in between the devices and substrate. A thin layer of thermally grown silicon dioxide (SiO_2) is an insulating layer which also work as masking layer for many other processes encountered during the device fabrication. A light-sensitive polymer (photoresist) is applied using spin coat to the wafer that can be either positive or negative depending on the process requirement. The wafer is exposed to ultraviolet (UV) light using the mask aligner, causing chemical changes in the photoresist. The exposed or unexposed regions of the photoresist are removed, depending on whether a positive or negative resist is used. Further, selective etching of the silicon oxide is carried out to open the window for the doping. In general, doping is carried out using ion implantation or diffusion for formation of N-well or P-well and then further n-type or p-type dopants are implanted/diffused into the source and drain regions of the transistors for making a CMOS devices (Figure 3.3). The wafer is heated to activate the dopants and repair any damage caused by the implantation process. Furthermore, if required a layer of polysilicon is deposited over the selective area of the wafer using CVD process. A layer of metal (typically aluminium) is deposited to form interconnections between devices which also requires photolithography and etching steps to define the metal interconnects. A protective layer generally silicon nitride or oxide is applied over the wafer to shield the devices from environmental damage that is further diced into individual chips and then packaging for integration into electronic systems that need to be conducted. In conclusion, the conventional processes and materials used in traditional IC technology are directly adopted for fabrication of 6G network devices. These processes are photolithography, thermal oxidation, dopant diffusion, ion-implantation, plasma activated chemical vapour deposition (PCVD), plasma-enhanced chemical vapour deposition (PECVD), evaporation, sputtering, etching, reactive ion etching (RIE), etc. Silicon, silicon dioxide, silicon nitride, polysilicon, aluminium are commonly used materials for the fabrication of IC and CMOS devices using these processes.

Figure 3.3 Process flow for CMOS-compatible fabrication of micro/nano devices.

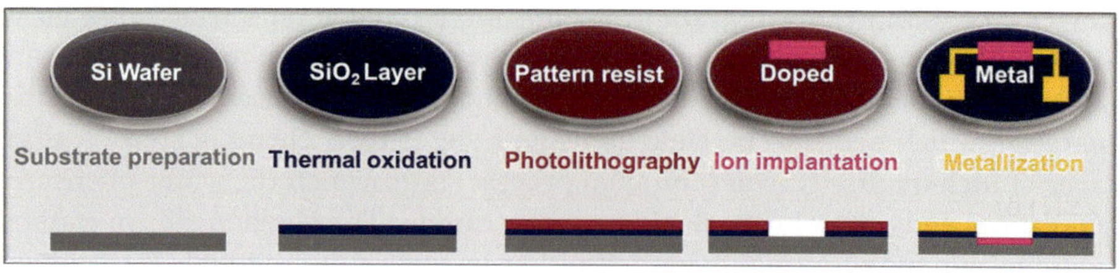

3.3.2 MEMS/NEMS techniques

Microelectromechanical systems (MEMS)-based sensors have become an integral part across a wide range of industrial and biomedical applications due to their compact size, high sensitivity, and integration capabilities. Various MEMS devices demonstrate this versatility. For instance, surface-micromachined parylene dual valves have been developed for on-chip, unpowered microflow regulation, offering precise control in microfluidic systems [6]. Similarly, compliant flexure diaphragms have been employed in micropumps, providing efficient fluid handling in biomedical and lab-on-chip platforms [6]. MEMS microswitches are also gaining traction in reconfigurable microwave circuits, enabling tuneable and adaptive RF systems. In the biomedical domain, nanoneedles serve as innovative electrode structures for biopotential measurements, supporting high-resolution bio-signal acquisition with minimal invasiveness. Furthermore, the integration of porous materials such as metal–organic frameworks (MOFs) with microelectronic systems is advancing applications in gas sensing, catalysis, and energy storage. MOFs offer high surface area and tuneable chemical properties, making them ideal for enhancing the functionality of MEMS-based sensing platforms [8]. Progress in hybrid manufacturing and assembly techniques has facilitated the development of three-dimensional multifunctional microelectronic systems. This includes the integration of MEMS with integrated circuits (ICs) at the chip scale. Mechanically guided 3D assembly approaches, which may exploit residual stresses or plastic deformation in metal films, allow for the creation of freestanding and complex 3D microstructures [8]. These techniques represent a paradigm shift in microsystem design and fabrication, moving beyond planar geometries. Energy management at the microscale is another critical area of advancement. Microgeneration technologies, which include the development of micro batteries and energy harvesting systems, are being explored to address the power demands of distributed MEMS devices. Innovative strategies such as energy crossover architectures with minimal battery capacity are under investigation for autonomous and long-term device operation [8]. On the nanoscale, nanoelectromechanical systems (NEMS) are emerging as a natural extension of MEMS, especially in applications requiring sub-10 nm resolution. NEMS devices benefit from ultra-small feature sizes and enhanced performance characteristics, making them suitable for high-frequency sensing and actuation.

To enable such nanoscale fabrication, a variety of scanning probe microscopy-based lithographic techniques have been developed. Tip-based nanofabrication (TBN) encompasses several methods, including scanning tunnelling microscope (STM)-assisted manipulation/lithography, thermal TBN, field-emission TBN, and dip pen nanolithography (DPN). These approaches allow for precise material deposition, patterning, and modification at the atomic and molecular levels, facilitating the fabrication of next-generation nanoscale devices.

Due to its ability to produce nanoscale patterns quickly and cost-effectively over large areas, nanoimprint lithography (NIL) is well-suited for several non-IC applications including micro- and nano-optics for fabrication of diffractive optical elements, microlens arrays, antireflective structures, photonic devices for patterning of photonic crystals and waveguides, plasmonic structuring of metal films at the nanoscale for enhanced electromagnetic field interactions, bio-nanotechnology for creation of nanostructured surfaces for biosensing and cell interaction studies, and energy devices for nano-patterning for thin-film solar cells and batteries to improve surface area and efficiency. Despite its numerous advantages, NIL faces several technical and practical challenges that limit its use in semiconductor IC production including defect sensitivity for physical contact between mold and resist can introduce particles and defects. It also poses overlay alignment issues for accurate multilayer patterning remains difficult due to alignment errors. Mold durability and cost are a few more limitation in view of fabricating large-area, high-resolution molds being expensive, and repeated use may degrade mold quality. Process reproducibility is a challenge for achieving uniform patterning over large substrates with minimal variation is non-trivial.

The traditional methods created for integrated circuit technology are used in the fabrication of silicon-based MEMS. For the creation of MEMS, three crucial steps from conventional silicon processing, (i) deposition, (ii) lithography, and (iii) etching, are used. Figure 3.4 provides a visual representation of such processes. Oxidation is one type of deposition; chemical vapour deposition, physical vapour deposition, and epitaxial growth are the other four types. Lithography comprises three types of beams: (i) optical, (ii) electron, and (iii) X-ray. Etching includes two methods: (i) dry etching and (ii) wet etching. For the creation of MEMS, many of these procedures are altered, such as deep reactive ion etching, greyscale lithography, and the use of thick photoresists. Techniques specifically developed for MEMS, not yet employed in IC fabrication, include surface micromachining, wafer bonding, electroplating, LIGA, ion beam etching, and deposition. To ensure long-term stability and reliability of microstructures, microsystems chips are bonded to another wafer, such as silicon, glass, metal, or polymeric substrates. The MEMS sensing chip is affixed to a glass substrate to create cavity sealing or vacuum, establish a reference cavity, or accommodate the deflection of a cantilever beam or thin diaphragm. The bonding methods for silicon wafers with other substrates include field-assisted bonding (anodic bonding) pioneered by Wallis and Pomerantz [9] and silicon fusion bonding (SFB) developed by Shimbo [10],

[11]. Field-assisted thermal bonding, encompassing anodic or electrostatic bonding, also known as the Mallory technique, is frequently employed to link silicon with glass at elevated temperatures (about 400 °C) and high voltages (around 600 V). Silicon fusion bonding (SFB) is a process to join two Si wafers at high temperature (>800 °C) in an oxidising environment without applying an electric field and no intermediate layer for bonding. Eutectic [12] and adhesive bonding [13] are employed for wafer bonding.

Figure 3.4 Schematic diagram showing the process of deposition, lithography and etching with both positive and negative photoresist [1].

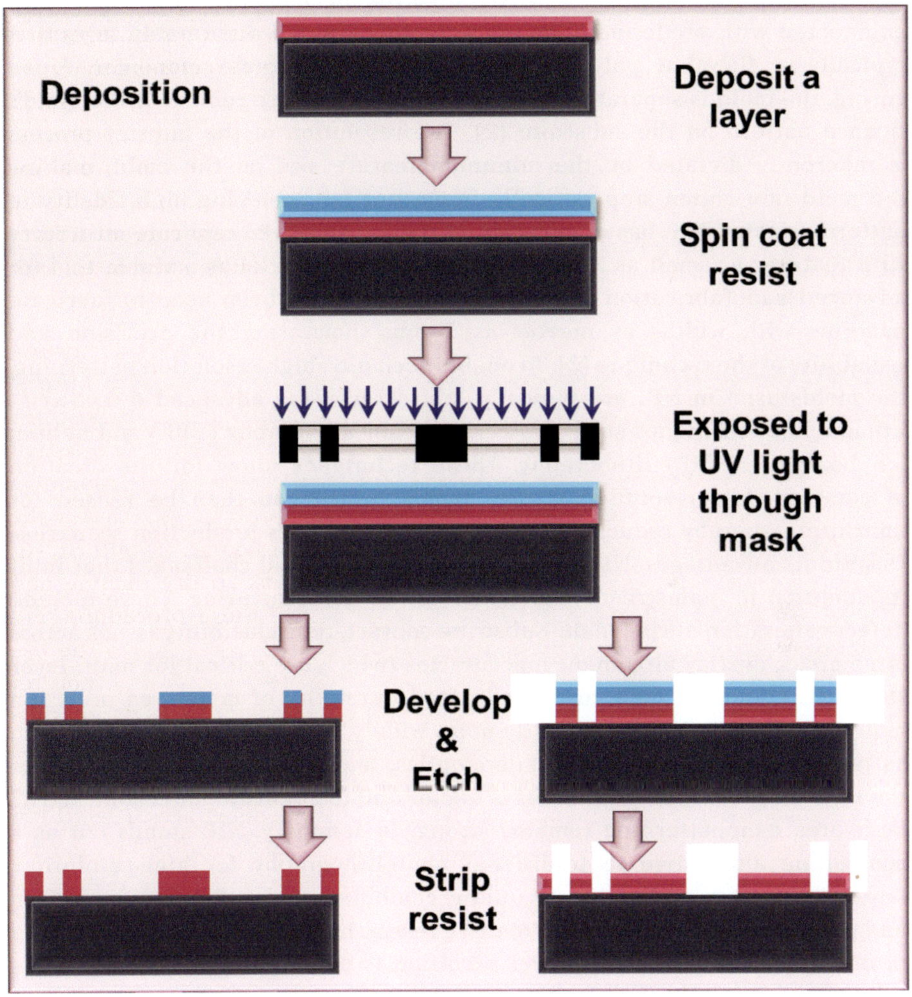

3.4 Dedicated Fabrication Processes

3.4.1 Nanoimprint and electron beam lithography

Nanoimprint lithography (NIL) is a scalable solution for high-resolution nanopatterning. NIL has emerged as a highly promising nanofabrication technique, particularly suited for large-scale production of nanoscale structures. Its relevance is especially pronounced in the fabrication of wafer-level optical components and other non-IC (integrated circuit) applications where high throughput, low cost, and sub-10 nm resolution are essential [3]. The fundamental process of NIL involves mechanically pressing a mold engineered with predefined nanoscale features onto a deformable substrate, typically a flowable polymer or a UV/thermal-curable monomer. Upon curing, the mold is separated, leaving behind an inverse replica of the mold's surface pattern on the substrate [2]. The resolution of the imprint process is inherently dictated by the minimum feature size on the mold, making the mold fabrication step critically important for achieving high fidelity in pattern transfer. NIL has demonstrated the capability to replicate structures with features as small as 2 nm, establishing its potential as a viable tool for advanced nanofabrication [2]. For instance, NIL has been used to fabricate nanofins with widths as narrow as 5 nm, showcasing the precision and scalability of the technique [2]. To enable such ultrahigh-resolution patterning, the molds used in NIL are typically fabricated using advanced direct-write lithographic techniques such as electron-beam lithography (EBL) and helium ion beam (He FIB) lithography. These techniques allow for the creation of complex, high-resolution master molds, which can then be reused for imprinting, thereby reducing the per-unit cost in mass production scenarios. Despite its advantages, NIL also faces several technical challenges that limit its adoption in mainstream semiconductor IC manufacturing. These include defect generation during mold–substrate contact, reproducibility issues across large areas, overlay alignment inaccuracies (which are critical for multi-layer structures), and the high cost and limited durability of large-area precision molds. As a result, NIL is currently more widely applied in non-IC fields such as nanophotonics, plasmonics, micro-optics, and bio-nanotechnology, where the tolerance for these limitations is higher and the demand for cost-effective, large-area nanopatterning remains strong. In summary, NIL stands out as a compelling alternative to traditional photolithography for high-resolution, large-area pattern replication. Its unique combination of scalability, resolution, and process simplicity continues to drive research and development, especially in emerging fields where IC-level precision is not a prerequisite. A process follow of process flow for nanoimprint lithography (NIL) for high-resolution nanopatterning is presented in Figure 3.5.

Figure 3.5 Process flow for nanoimprint lithography (NIL) for high-resolution nanopatterning.

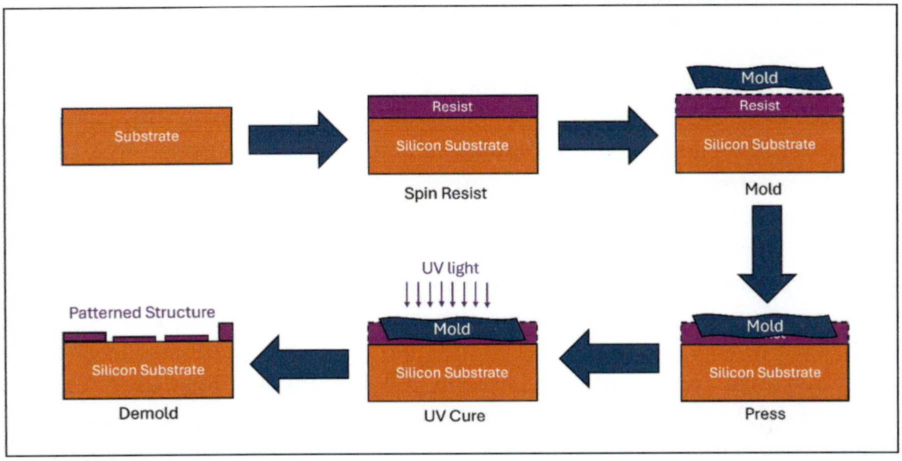

3.4.2 Electron beam lithography (EBL)

Electron beam lithography (EBL) is a precision tool for micro/nanoscale fabrication. EBL is a high-resolution, maskless lithography technique widely used in nanofabrication to define patterns with feature sizes well below the diffraction limits of conventional photolithography. Renowned for its capability to achieve sub-10 nm resolution, EBL is considered indispensable for applications requiring extreme precision, such as nanophotonic, quantum devices, and next generation metasurfaces [2]. The working principle of the core capabilities of EBL operates by focusing a highly accelerated beam of electrons (typically 10–100 keV) onto an electron-sensitive resist (such as PMMA or ZEP) coated on a substrate. As the electron beam scans across the resist according to a predefined CAD pattern, localized exposure occurs, altering the solubility of the resist in a developer solution. This enables either positive or negative patterning, depending on the resist type and development process. One of EBL's defining strengths is its ultra-high resolution: patterns with sub-10 nm features have been routinely demonstrated, which makes it highly suitable for fabricating master molds for nanoimprint lithography (NIL) [3].

Its ability to directly write on a resist without requiring a physical mask also offers unmatched flexibility for prototyping and low-volume manufacturing. Beyond its resolution, EBL has proven effective in patterning a wide range of 2D materials and nanostructures. For instance, graphene patterning over large areas with minimum feature sizes as small as 350 nm has been achieved using EBL, facilitating its integration into nanoelectronic and optoelectronic devices [3]. EBL has also been used to define intricate square patterns, a critical step in tailoring the optical, electronic, and mechanical properties of transition metal dichalcogenide (TMD) for device fabrication [3]. These capabilities highlight EBL's role in the precise manipulation of atomically thin materials. One of the more advanced capabilities of EBL is greyscale lithography, where variable electron beam doses are applied to create topographic features of varying depths within a single patterning step [5]. This is particularly useful in applications such as micro-optics, where depth-graded features are used to create lenses, holograms, or diffraction gratings. Moreover, EBL serves as a central tool in multistep lithography workflows, especially for the fabrication of complex or hybrid nanostructures such as metasurfaces. These structures often require precise layering and alignment of nanoscale features across multiple fabrication steps. EBL is frequently integrated with other techniques, including reactive ion etching (RIE), for anisotropic etching and pattern transfer, photolithography for large-area or coarse features in hybrid processes, lift-off processes for defining metal nanostructures after patterning, and deposition. This integration enables the realization of functional devices that rely on both nanoscale detail and large-area uniformity. EBL is widely employed in metasurface fabrication for light manipulation, polarization control, and wavefront engineering. It is also used in quantum dot and nanowire patterning for next-generation electronics and photonics. However, EBL is not without limitations. The process is serial and time-consuming, leading to low throughput, which restricts its use in mass production. Additionally, the equipment cost, and maintenance requirements are high, and proximity effects caused by electron scattering within the resist and substrate can affect pattern fidelity if not properly corrected. Electron beam lithography remains a cornerstone technique in the nanofabrication landscape, offering unmatched resolution and versatility. Whether used for prototyping, mold fabrication, or as part of hybrid lithographic strategies, EBL continues to enable innovations at the frontiers of nanoscience and nanotechnology. A schematic of the process flow for electron beam lithography (EBL) for high-resolution nanopatterning is presented in Figure 3.6.

Figure 3.6 Process flow for electron beam lithography (EBL) for high-resolution nanopatterning.

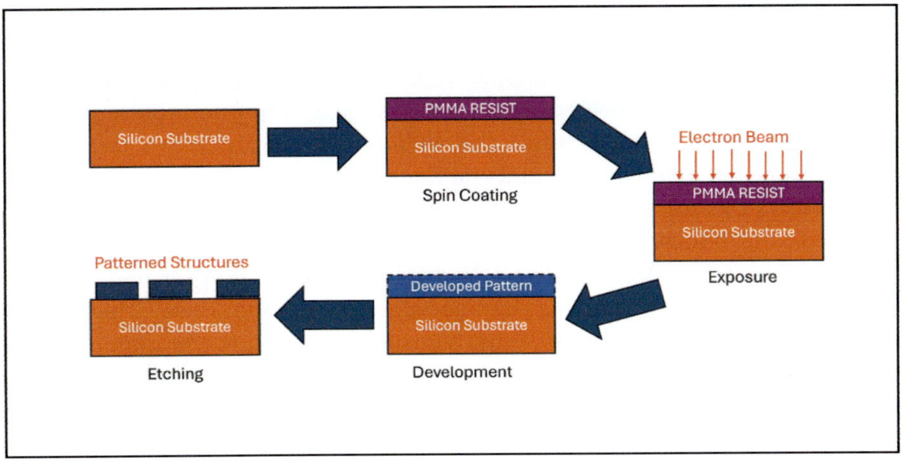

3.4.3 Focused ion beam (FIB)

Focused ion beam (FIB) micro/nanofabrication is another precision tool for 3D and sub-10 nm structuring of micro/nano devices. FIB systems have become vital tools in nanofabrication and prototyping, offering capabilities that extend beyond conventional lithographic techniques. FIB employs a finely focused beam of ions, most commonly gallium (Ga^+) or helium (He^+), to directly interact with materials at the nanoscale, enabling a broad range of processes including direct-write lithography, milling, ion-induced deposition, and localized etching [2]. The advent of helium ion beam lithography (He^+-FIB) has pushed the boundaries of resolution in direct-write nanofabrication for sub-10 nm patterning. Due to the shorter de Broglie wavelength and smaller interaction volume of helium ions compared to electrons or heavier ions, He^+-FIB systems can produce features below 10 nm with excellent edge definition and minimal proximity effects. For example, the fabrication of nested L patterns with a pitch of just 10 nm has been demonstrated using He^+-FIB, showcasing its extreme precision and suitability for dense nanostructuring [2]. Moreover, sub-5 nm patterning using FIB has been successfully achieved, positioning it as a serious contender for applications where ultra-fine resolution is critical and mask-based approaches are impractical [2]. FIB's most prominent strength lies in its material-removal (milling) capabilities, which allow direct sculpting of materials in three dimensions. This has led to its widespread use in the fabrication of complex 3D nanostructures and functional devices, such as nanocantilevers for sensing applications. Self-rolled multilayer metasurfaces, where FIB is used to define release patterns for strain-induced rolling, chiral and helical 3D architectures, critical for polarization-sensitive optics and metamaterials [4]. These applications underscore FIB's unmatched versatility for direct and localized modification of micro- and nanoscale structures. In addition to binary lithography, FIB milling serves as an alternative to EBL for greyscale nanopatterning, offering depth-controllable structuring by modulating the ion dose. This allows the creation of voids or reliefs with continuous variations in diameter and depth, enabling applications in micro-optics, holography, and advanced metasurfaces [4]. The direct-write capability of FIB makes it highly suitable for rapid prototyping or fabricating one-off custom designs with varying 3D profiles. Despite its powerful capabilities, FIB technology faces several practical limitations that constrain its use in large-scale nanofabrication. High cost and infrastructure requirements are among a few limitations of FIB systems as these are expensive to acquire and maintain, limiting accessibility to specialized research facilities. Since FIB is a serial technique (like EBL), patterning even moderately large areas is time-intensive, extremely low throughput, and unsuitable for high-volume manufacturing (Figure 3.7). While FIB provides excellent resolution, maintaining precise feature geometry across complex or deep structures remains a challenge. The vertical resolution is often limited by ion beam scattering and redeposition effects, affecting the aspect ratios achievable in deep etching or milling. High-energy ions can cause lattice damage, amorphization, or implantation into sensitive materials, which is especially critical for semiconductor or biological samples [4]. Focused ion beam technology stands out as a versatile and powerful nanofabrication technique, capable of generating sub-5 nm features and intricate 3D geometries. While limited by throughput and material damage, its direct-write, maskless approach makes it indispensable for research, device prototyping, and advanced structuring tasks that cannot be accomplished using other techniques.

3.4.4 2D material integration

Fabrication techniques for the on-chip integration of 2D materials along with their advantages and drawbacks are summarised and discussed in the source [3]. These techniques are categorised into material synthesis, on-chip transfer, film patterning, and property tuning/ modification. Emerging techniques to implement 2D van der Waals heterostructures on chips are also reviewed [3]. 2D materials such as graphene, transition metal dichalcogenides (TMDCs), black phosphorus (BP), graphene oxide (GO), perovskites, MXenes, and metal–organic frameworks (MOFs) have been used in applications like field-effect transistors (FETs) and compact devices[3]. Figure 3.8 presents a workflow for on-chip integration of 2D materials like graphene and MoS_2 for 6G devices.

Figure 3.7 Process flow for focused ion beam microfabrication.

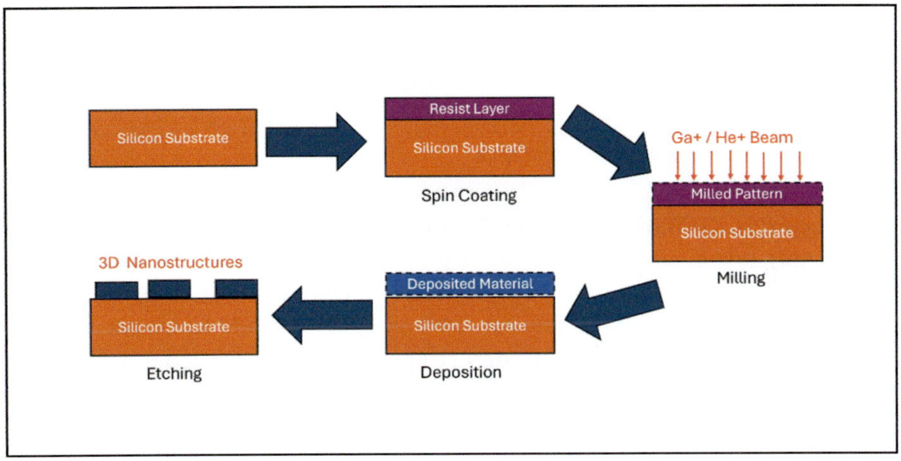

Figure 3.8 Workflow for on-chip integration of 2D materials like graphene and MoS_2 for 6G devices.

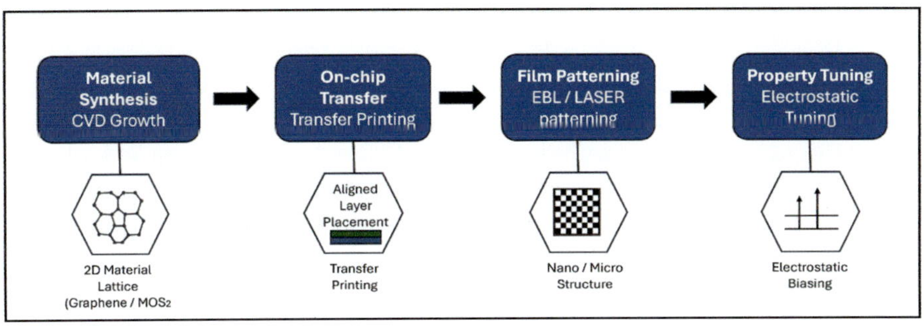

Film patterning of 2D materials can be achieved through various methods:

- **Laser patterning**, including direct laser writing (DLW), can create features like graphene nanoribbons, graphene nanodisk arrays, MoS_2 nanohole arrays, WSe_2 lenses, GO microcircuits, rGO flat lenses, stretchable GO metalenses, micropatterns on BP flakes, MXene microcircuits, and perovskite patterns [3]. Laser-induced techniques can also process complex amplitude-modulating metasurface devices using laser-induced graphene [3]. Laser-based nanomanufacturing is a versatile and mask-free technique [4].
- **Lithography patterning**, such as EBL and ultraviolet lithography, can fabricate large-area graphene patterns with feature sizes down to 350 nm [3]. EBL can pattern MoS_2 monolayers [1]. Lithography and etching processes are used [3].
- **Nanoimprinting** techniques can also be used for 2D film patterning, producing structures like patterned graphene nanoring arrays, graphene discs, and graphene nanomesh [3].
- **Transfer printing** is discussed as a method, including its integration with post-assembly approaches to fabricate sub-10 nm metallic nanogaps on MoS_2 flakes [1].

3.4.5 Metasurface and antenna fabrication

Metamaterials (MMs) are artificially manufactured materials composed of periodically arranged unit cell structures, typically with minimal feature sizes in the sub-wavelength range [5]. Electromagnetic MMs, such as absorbers, antennas, lenses, and filters, control electromagnetic fields and waves and are used in applications like stealth, detection, imaging, and communication [5]. Metasurfaces (MSs) comprise periodic sub-wavelength nano antennas and are studied using various materials like TiO_2, perovskite, and silicon [4].

Fabrication techniques for metasurfaces include:

(1) Standard nanolithography techniques like EBL, FIB, and photolithography [14].
(2) Using EBL to form reverse nanofins in resist, filling with amorphous TiO_2 through Atomic Layer Deposition (ALD), followed by blanket RIE [4].
(3) Fabricating TiO_2 films using E-beam evaporation [4].
(4) Using optimised FIB parameters for creating high-quality structures, although deviations can increase surface roughness and decrease precision [4]. FIB is also used for 3D nanostructures [4].
(5) Laser-based nanomanufacturing, including patterned pulse laser direct writing (LDW) lithography for sub-wavelength structural units [4]. One-step LDW based on laser-induced techniques can create complex amplitude-modulating metasurfaces [4].
(6) Advanced lithography methods like greyscale lithography, which allows varying light intensity during exposure to create nanostructures with different heights or depths [4]. EBL and FIB milling can be used for greyscale nanopatterning [4].

(7) Multistep lithography, involving a sequence of lithographic processes (e.g., combining EBL with laser beam lithography or photolithography) to create intricate nanostructures or hybrid metasurfaces with metal and dielectric materials [4]. Two-step EBL can fabricate multilayer metasurfaces or hybrid metal–dielectric structures[4] .

(8) Nanoimprint lithography (NIL) can also be used, for example, imprinting a solution of TiO_2 nanocrystal[4] .

(9) Using deep Bosch processes to etch high-aspect-ratio structures, such as ultradeep via-holes or silicon micropillars for THz waves [4].

3.4.6 Metamaterial absorber fabrication

The fabrication of metamaterial absorbers (MMAs) has traditionally relied on a suite of well-established microfabrication and thin-film deposition techniques. These include printed circuit board (PCB) manufacturing, photolithography, magnetron sputtering, chemical vapor deposition (CVD), and micro/nanomechanical processing methods [5]. Such processes have enabled the creation of various planar and multilayered absorber structures, often with precise control over thickness, layer composition, and surface patterning. However, despite their proven utility, these conventional approaches face significant challenges particularly when tasked with producing complex three-dimensional (3D) geometries, integrating multiple functional materials, or achieving rapid prototyping. Limitations in material compatibility, structural design freedom, and multi-step process complexity restrict their adaptability for next-generation MMA architectures [5]. To address these shortcomings, 3D printing, or additive manufacturing (AM), has emerged as a promising alternative fabrication strategy for MMAs. Unlike subtractive techniques, AM offers layer-by-layer freeform fabrication, enabling the creation of highly intricate 3D structures that are otherwise difficult or impossible to produce using traditional methods. The inherent flexibility of 3D printing also supports the integration of diverse materials, including polymers, ceramics, and conductive inks, thereby broadening the design space and functional scope of MMAs[5].

Among the various 3D printing technologies employed in MMA fabrication, three primary process categories stand out:

(1) **Material extrusion (e.g., fused deposition modelling, FDM):** In this method, thermoplastic or composite filaments are extruded through a heated nozzle to build up structures layer by layer. It is particularly suited for producing macro-scale absorbers with embedded conductive or dielectric patterns.

(2) **Vat photopolymerization (e.g., stereolithography, SLA and digital light processing, DLP):** These techniques utilize photosensitive resins cured by a light source to achieve high-resolution 3D structures. SLA and DLP are especially effective for printing complex lattice geometries or finely detailed MMA units, making them ideal for applications requiring precision and smooth surface finishes.

(3) **Powder bed fusion (e.g., selective laser sintering, SLS and direct metal laser sintering, DMLS):** Powder-based methods enable the fabrication of MMAs using metallic or ceramic materials, offering superior mechanical robustness and thermal stability, essential for high-frequency or harsh environment applications.

Notable examples of 3D-printed MMAs include structures with twisted cross geometries and cross-helix unit cell configurations, which are designed to achieve tailored absorption across specific frequency bands [5]. These complex unit cells leverage the design freedom of 3D printing to tune electromagnetic responses through shape anisotropy, coupling effects, and sub-wavelength structuring capabilities that are particularly relevant for applications in THz sensing, stealth technology, and broadband EM shielding. While traditional MMA fabrication techniques remain relevant for planar and multilayer designs, 3D printing technologies represent a transformative shift, enabling the realization of highly customized, structurally complex, and functionally diverse MMAs. As material libraries expand and printer resolution continues to improve, additive manufacturing is poised to play an increasingly central role in the next generation of adaptive, conformal, and high-performance metamaterial absorbers.

3.5 Hybrid Fabrication Approaches

The development of three-dimensional (3D) multifunctional micro/nano electronic systems is increasingly being driven by hybrid manufacturing and assembly strategies, which combine multiple fabrication methodologies to overcome the limitations of single-process approaches [8]. These hybrid methods are particularly valuable for realizing devices that require the integration of heterogeneous components and materials, such as electronic, photonic, and mechanical elements within a unified architecture. A notable example is the chip-scale integration of micro-electro-mechanical systems (MEMS) with integrated circuits (ICs), which demands precise alignment, interconnection, and compatibility between differing fabrication technologies and materials [8]. Additionally, mechanically guided 3D assembly techniques, when combined with residual stress-based deformation, plastic buckling, or soft lithographic transfer processes, can transform 2D precursor structures into complex freestanding 3D architectures, enabling the creation of spatially reconfigurable or multifunctional devices [8]. In the context of nanophotonic and electromagnetic device engineering, hybrid patterning techniques are pivotal for fabricating advanced metasurfaces with intricate and diverse nanostructures [4]. These approaches often combine multiple lithography methods, such as electron beam lithography (EBL), nanoimprint lithography (NIL), and greyscale photolithography, to achieve high-resolution features and tailored functionalities across different material systems. For instance, hybrid metasurfaces composed of metallic nanostructures (e.g., gold nanorods) coupled with dielectric resonators (e.g., silicon nanodisks) have been fabricated using two-step EBL processes, enabling enhanced spectral control and polarization responses [4]. Multilayer metasurfaces further expand functionality by stacking different patterned layers, each contributing specific optical or electromagnetic properties (Figure 3.9).

Figure 3.9 Schematic of hybrid metasurface fabrication combining EBL and NIL for 6G applications.

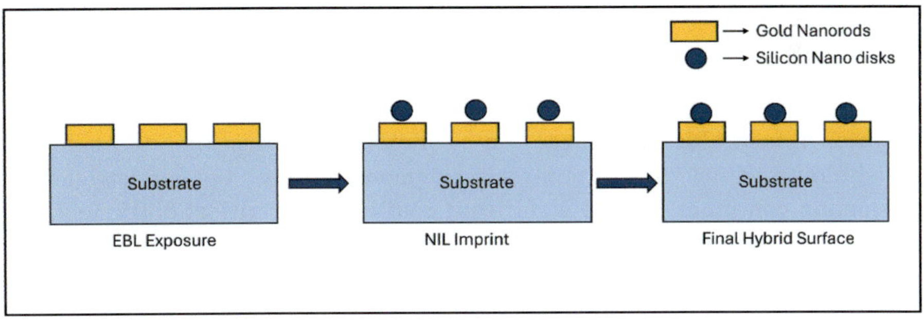

Fabrication techniques for such architectures often involve combinations of lithography, reactive ion etching (RIE), and encapsulation processes such as polydimethylsiloxane (PDMS) transfer and lamination [4]. More recently, multiphoton lithography (MPL), a form of advanced 3D printing based on two-photon polymerization has emerged as a powerful tool to directly fabricate multilayer, freeform nanostructures with sub-diffraction resolution, bypassing the need for multiple etching and alignment steps [4]. An illustrative application includes the fabrication of hybrid achromatic metalenses, where a phase compensation plate and a focusing metalens are monolithically integrated using MPL to correct chromatic aberrations over broadband wavelengths [5].

Another innovative hybrid approach involves transfer printing combined with post-assembly mechanical methods to fabricate ultrasmall nanogaps, which are critical for applications in plasmonics, biosensing, and tunnelling-based electronics [1]. This technique entails pre-patterning nanostructures on a stretchable substrate, followed by mechanical deformation (stretching) and subsequent release, causing the transferred components to move closer and form sub-10 nm gaps due to elastic relaxation [1]. Such strategies provide deterministic control over nanoscale spacing without relying on expensive lithographic alignment tools.

Together, these hybrid fabrication and assembly strategies represent a transformative direction for next-generation device manufacturing, offering new capabilities for building complex, high-performance micro- and nanosystems. They enable the realization of integrated platforms that simultaneously leverage the unique advantages of different material classes, geometrical configurations, and processing techniques that are a necessity for applications ranging from 6G communication devices and tunable optics to quantum sensors and wearable electronics.

3.6 Challenges and Future Directions

3.6.1 Scalability and yield

Despite remarkable advances in nanoscale fabrication technologies, transitioning from laboratory-scale demonstrations to industrial-scale production continues to pose significant challenges. A critical bottleneck lies in achieving high yield and scalability while maintaining pattern fidelity and functional integrity. Nanoimprint lithography (NIL), although promising for large-area, low-cost patterning, still struggles with the cost-effective fabrication of durable molds that can support repeated use without degradation. Ensuring uniform, high-fidelity pattern transfer across entire wafers or substrates remains a complex engineering task [2]. Similarly, focused ion beam (FIB) lithography, while offering unmatched resolution and 3D patterning capabilities, suffers from extremely low throughput, rendering it impractical for high-volume manufacturing scenarios [4]. Traditional micro-mechanical assembly methods face their own limitations, particularly in realizing intricate or multiscale 3D structures. These methods often require complex, multistep procedures that are both time-consuming and inflexible [5]. In contrast, 3D printing technologies, including direct laser writing and two-photon polymerization, present a promising solution to these limitations. They enable the streamlined production of 3D micro- and nanostructures with considerable design freedom and can potentially enhance throughput and reduce process complexity [5]. Furthermore, block copolymer (BCP) self-assembly has emerged as a scalable, bottom-up approach for patterning nanoscale features. Its inherent simplicity, low cost, and suitability for large-area processing make it a compelling option for high-throughput device fabrication, especially in areas such as nanoelectronics and soft lithography[1].

3.6.2 Heterogeneous integration

The next frontier in micro- and nanodevice fabrication lies in the heterogeneous integration of multiple functional materials, components, and subsystems into compact 3D architectures. This is particularly important for applications in multifunctional microelectronics, where different material systems (e.g., metals, semiconductors, dielectrics, polymers) must be integrated at the microscale. Examples include chip-scale integration of MEMS and ICs, where mechanical and electrical domains coexist within a unified platform [8], and the fabrication of hybrid metasurfaces, which often involve the combination of metallic and dielectric nanostructures to achieve novel optical functionalities [4]. The realization of such integrated systems requires the interfacing of distinct fabrication methods such as top-down lithography with bottom-up self-assembly, or additive manufacturing with subtractive techniques and demands robust process compatibility, interface engineering, and precise alignment across scales. As device complexity grows, process interoperability and material compatibility will become increasingly crucial to the successful deployment of next-generation systems.

3.6.3 Cost and environmental considerations

The economic and environmental impacts of nanofabrication methods are becoming increasingly relevant, particularly in the context of sustainable manufacturing and emerging applications where CMOS-based processes may be cost-prohibitive[2].

Techniques such as extreme ultraviolet (EUV) and deep ultraviolet (DUV) lithography offer exceptional resolution, but their high capital and operational costs restrict accessibility to only the most advanced semiconductor fabs [1]. FIB systems, despite their versatility, are also associated with high equipment and maintenance costs, further limiting their widespread adoption [4]. In contrast, methods like laser direct writing (LDW) and 3D printing are noted for their cost-efficiency and reduced processing complexity [4][5]. LDW allows for maskless, flexible patterning at relatively low expense, while additive manufacturing eliminates several lithography and etching steps, reducing both fabrication time and material waste. Although environmental considerations are not always explicitly discussed, several emerging approaches such as BCP self-assembly, NIL, and 3D printing implicitly offer greener alternatives due to their lower energy usage, reduced chemical processing, and potential for process integration. As sustainability becomes a global priority, eco-friendly materials, solvent-free patterning, and energy-efficient processing will be vital research directions.

Addressing the intertwined challenges of scalability, integration complexity, cost, and sustainability is essential for the continued evolution of nanofabrication technologies. The future will likely involve hybrid approaches that combine the strengths of various techniques merging precision with efficiency, and scalability with multifunctionality to meet the demands of next-generation electronics, photonics, and biomedical devices.

3.7 Conclusion

This chapter has examined the critical role of micro- and nanofabrication technologies in enabling the next generation of electronic and photonic devices, with a focus on their relevance to 6G networks and beyond. Drawing from a range of fabrication methodologies, we have traced the evolution from traditional techniques to emerging hybrid approaches tailored to meet the performance, scalability, and integration demands of future systems. Traditional CMOS-based fabrication techniques have demonstrated impressive capabilities, achieving sub-10 nm resolution, thereby forming the technological foundation of contemporary integrated circuits. However, their high capital cost, process complexity, and material compatibility constraints present significant barriers for emerging applications outside the conventional semiconductor industry [3]. These limitations have prompted the exploration of alternative and complementary fabrication techniques capable of offering greater flexibility, cost-effectiveness, and integration with novel materials.

The advent of 6G technologies, particularly those operating in the terahertz (THz) frequency regime, has introduced a new set of requirements for components such as reconfigurable metasurfaces, ultra-thin antenna arrays, and high-frequency modulators. These devices often demand nanoscale precision and involve advanced materials like graphene and other 2D materials, which exhibit extraordinary electronic, optical, and thermal properties [6]. Successfully integrating these materials requires specialized processing techniques, including chemical vapor deposition (CVD), transfer printing, laser patterning, nanoimprinting, and high-resolution lithography [3].

Key fabrication techniques discussed include:

(1) Electron beam lithography (EBL), which offers unparalleled resolution for direct-write nanopatterning and is particularly effective in prototyping and greyscale lithography applications [4];
(2) Focused ion beam (FIB) methods, which support not only direct patterning but also milling, deposition, and etching for 3D nanostructures, albeit with limitations in throughput and cost [3,4].
(3) Nanoimprint lithography (NIL), which enables sub-10 nm pattern replication and is considered a cost-effective method for large-area, non-IC applications despite challenges in mold fabrication and durability[2].

The fabrication of metasurfaces and metamaterial absorbers, which are crucial for THz beam shaping and tunability, often employs a combination of standard photolithography, laser-based patterning, multistep greyscale methods, and increasingly, 3D micro- and nanoscale printing. These approaches facilitate the creation of complex, multi-layered, and reconfigurable structures, enabling novel electromagnetic responses in a compact footprint [4].

To meet the increasing demand for multifunctional and heterogeneous integration, hybrid fabrication approaches are emerging as essential tools. These integrate top-down and bottom-up techniques, and often involve combinations of MEMS, ICs, 2D materials, and photonic components, thereby allowing the realization of compact, high-performance, and adaptive devices for future networked systems [8].

Despite these advances, key challenges persist. Scalability, throughput, process repeatability, and cost-efficiency remain major concerns, especially for transitioning laboratory-scale techniques into mass production environments. The high equipment costs associated with tools like EBL and FIB, the limited lifetime and reproducibility issues in NIL, and the precision alignment needs in heterogeneous integration all demand focused innovation [2,3,4].

The way forward for future research and development efforts will likely prioritize:

(1) Scalable, low-cost manufacturing solutions capable of maintaining nanoscale precision,
(2) Eco-friendly and sustainable fabrication techniques that reduce environmental impact,
(3) Process standardization and automation for hybrid systems integration,
(4) Enhanced material-specific patterning techniques for next-generation quantum, optoelectronic, and THz devices.

Ultimately, the convergence of advanced materials, innovative fabrication strategies, and scalable manufacturing paradigms will be critical in realizing the vision of 6G networks and intelligent, adaptive, nanoscale device ecosystems.

References

[1] H. Kang et al., "Emerging low-cost, large-scale photonic platforms with soft lithography and self-assembly," May 22, 2023, SPIE. doi: 10.3788/PI.2023.R04.

[2] Y. Chen et al., "Sub-10 nm fabrication: Methods and applications," Jul. 01, 2021, IOP Publishing Ltd. doi: 10.1088/2631-7990/ac087c.

[3] D. J. Moss, "Fabrication methods for integrating 2D materials."

[4] W. Yang, J. Zhou, D. P. Tsai, and S. Xiao, "Advanced manufacturing of dielectric metadevices," Jun. 30, 2024, SPIE. doi: 10.3788/PI.2024.R04.

[5] J. Peng et al., "Review of micro and nano scale 3D printing of electromagnetic metamaterial absorbers: mechanism, fabrication, and functionality," 2024, Taylor and Francis Ltd. doi: 10.1080/17452759.2024.2378937.

[6] H. Taghvaee et al., "Multiwideband Terahertz Communications Via Tunable Graphene-Based Metasurfaces in 6G Networks: Graphene Enables Ultimate Multiwideband THz Wavefront Control," IEEE Vehicular Technology Magazine, vol. 17, no. 2, pp. 16–25, Jun. 2022, doi: 10.1109/MVT.2022.3155905.

[7] A. Shafie, N. Yang, C. Han, J. M. Jornet, M. Juntti, and T. Kurner, "Terahertz Communications for 6G and Beyond Wireless Networks: Challenges, Key Advancements, and Opportunities," Jul. 2022, [Online]. Available: http://arxiv.org/abs/2207.11021

[8] N. Singh et al., "Challenges and opportunities in engineering next-generation 3D microelectronic devices: improved performance and higher integration density," Sep. 26, 2024, Royal Society of Chemistry. doi: 10.1039/d4na00578c.

[9] G. Wallis, "Field assisted glass-metal sealing," J Appl Phys, vol. 40, no. 10, pp. 3946–3949, 1969, doi: 10.1063/1.1657121.

[10] M. Shimbo, K. Furukawa, K. Fukuda, and K. Tanzawa, "Silicon-to-silicon direct bonding method," J Appl Phys, vol. 60, no. 8, pp. 2987–2989, 1986.

[11] P. W. Barth, "Silicon fusion bonding for fabrication of sensors, actuators and microstructures," Sens Actuators A Phys, vol. 23, no. 1–3, pp. 919–926, Apr. 1990, doi: 10.1016/0924-4247(90)87060-V.

[12] M. B. Cohn and R. T. Howe, "Wafer-to-wafer transfer of microstructures using break-away tethers," 6,142,358, Nov. 2000

[13] D. Maas et al., "Fabrication of microcomponents using adhesive bonding techniques," Proceedings of Ninth International Workshop on Micro Electromechanical Systems, pp. 331–336, 1996, doi: 10.1109/MEMSYS.1996.494003.

[14] M. Manvi and K. B. Mruthyunjaya Swamy, "Microelectronic materials, microfabrication processes, micromechanical structural configuration based stiffness evaluation in MEMS: A review," Jul. 15, 2022, *Elsevier B.V.* doi: 10.1016/j.mee.2022.111854.

CHAPTER

4

Federated Multi-agent Reinforcement Learning for Edge-intelligent Beamforming Codebook Design for mmWave MIMO

Manasjyoti Bhuyan[1], Kandarpa Kumar Sarma[1], Purabi Sharma[1], Aradhana Misra[1], Debashis Dev Misra[2], Jacopo Iannacci[3] and Koushik Guha[4]

[1]Department of Electronics and Communication Engineering, Gauhati University, Guwahati- 781014, Assam, India
[2]Department of Computer Science and Engineering, Assam down town University, Guwahati, Assam, India
[3]Center for Sensors and Devices (SD), Fondazione Bruno Kessler, Povo, 38123 Trento, Italy
[4]Department of Electronics and Communication Engineering, NIT Silchar, Silchar, Assam, India

Email: manasjyoti.b@gmail.com (M.B); kandarpaks@gauhati.ac.in (K.K.S); purabis1989@gmail.com (P.S); aradhana.misra@gauhati.ac.in (A.M); debashish.dm@gmail.com (D.D.M); iannacci@fbk.eu (J.I); koushik@ece.nits.ac.in (K.G.)

Abstract

Hybrid beamforming with massive multiple-input multiple-output (MIMO) antenna arrays plays a pivotal role in achieving high data rates and spectral efficiency in next-generation wireless systems. Nevertheless, the real-time

DOI: 10.1201/9788743809197-4

design and adaptation of beamforming codebooks and user grouping remain challenging, as conventional machine learning (ML) methods typically rely on the centralized collection of extensive channel state information (CSI), resulting in substantial feedback overhead and potential privacy violations. This chapter proposes a federated multi-agent reinforcement learning (MARL) framework for joint hybrid beamforming and dynamic user grouping. Unlike prior work which focuses solely on centralized codebook learning, the proposed approach enables distributed user equipment (UE) to collaboratively train codebook and clustering algorithms without sharing raw data, thereby preserving privacy. The proposed method significantly reduces communication overhead while achieving near-optimal beam patterns and robust scalability in dense network scenarios, making it well-suited for practical deployment in next-generation wireless systems. The experimental findings indicate that for mobile users, federated learning (FL)-based user clustering with MARL outperforms traditional initial access-based beam clustering in MARL.

4.1 Introduction

The next generation of wireless networks (beyond 5G/6G) is expected to leverage advanced signal processing and machine learning (ML) to meet ever-growing demands for capacity, coverage, and latency. In particular, hybrid (analog–digital) beamforming techniques with massive multiple-input multiple-output (MIMO) antenna arrays are key enablers of high data rates and spectrum efficiency. However, designing and adapting beamforming codebooks and user groupings in real time is challenging; traditional ML approaches require collecting large channel state information (CSI) datasets at a central server, leading to prohibitive feedback overhead and privacy concerns. Recent surveys note that the ever-increasing data volume, along with users' privacy requirements and the rising computing power at the network edge, are driving a paradigm shift toward decentralized learning in wireless systems. Federated learning (FL) and distributed multi-agent reinforcement learning (MARL) have emerged as promising frameworks to address these challenges. In this context, user devices themselves can participate in training ML models without exposing raw data, while coordinating with the base station (BS) to improve global performance.

4.1.1 Edge intelligence on modern mobile devices

Modern smartphones and IoT devices are no longer resource-starved; they often include powerful GPUs, NPUs, and dedicated AI accelerators (e.g., Qualcomm AI engines, Apple Core ML, Huawei NPUs), enabling on-device deep learning inference and even local training. For example, the Huawei Kirin 970

SoC incorporates a neural processing unit that can classify 2000 images per second on-device. These hardware advances, combined with lightweight mobile ML frameworks (TensorFlow Lite, Core ML, etc.), allow user equipment (UE) to execute and train neural networks directly. Simultaneously, user devices generate rich sensory data, including wireless CSI estimated from pilots, inertial measurements unit (IMU), global positioning system (GPS) location, and traffic patterns, which are valuable for learning. Edge learning literature emphasizes that bringing computation to the data can substantially reduce latency and backhaul load, while addressing privacy and bandwidth constraints. In other words, on-device learning is now feasible; small reinforcement learning (RL) agents can be trained locally on each device using its own CSI, location, and motion data as observations. These local agents can learn to adjust transmit/receive beam selections, pilot allocations, or other radio parameters based on instantaneous context, without requiring raw data to leave the device.

4.1.2 Federated learning for confidential collaboration

FL provides a natural mechanism to collaboratively train global models across edge devices while keeping raw data local. In FL, each UE computes updates on its local data and only the model updates are sent to the BS for aggregation. This *train-at-edge* approach was originally motivated by privacy-sensitive mobile data. It should be noted that modern devices hold rich user data (language, images, etc.) that cannot be centralized for training. FL circumvents this by *leaving training data on-device and only sharing model updates*, yielding strong data confidentiality guarantees. At the same time, FL is communication-efficient; experiments show it can reduce the number of communication rounds by 10–100 times compared to naive synchronous SGD. In wireless networks, applying FL means that CSI or sensor measurements of each user are never directly exposed, instead, the BS learns from aggregated signals that summarize many devices' experiences. This preserves user privacy and limits fronthaul load. For example, in federated hybrid beamforming, users only send neural network gradients (not full CSI matrices) to the BS, dramatically lowering feedback overhead. FL also inherently adapts to heterogeneous data and device capabilities, making it suitable for non- independently and identically distributed (non-i.i.d) wireless channel conditions. Overall, FL is an enabling technology for confidential, distributed learning in the radio domain.

4.1.3 Synergy of federated learning and multi-agent RL

MARL extends single-agent RL to cooperative settings where multiple decision-makers (agents) interact. In a wireless system, each UE or each access point

(AP) can act as an RL agent making localized decisions based on its own state and reward. Integrating MARL with FL yields a *federated MARL* framework; each device trains its local policy network on-device, and the base station periodically aggregates policy parameters or value-function updates across devices. This approach leverages the decentralized and collaborative aspects of MARL while exploiting FL's communication/privacy advantages. Recent work observes that wireless networks and MARL have complementary capabilities. The decentralized topology of cell-free or IoT systems aligns naturally with multiple agents, and coordinated learning can unlock 6G performance goals.

Federated MARL has been shown to be a significant improvement over purely centralized or isolated learning. For example, in cell-free massive MIMO, Banerjee et al. developed a MARL algorithm in which each AP (agent) learns which users to serve [1]. By adding a federated training step, they obtain a shared clustering policy that generalizes across environments: both conventional and federated MARL outperformed baseline AP-selection schemes, with the federated version particularly robust to environmental changes. Similarly, DRL-based codebook design methods use neural networks and clustering of user CSI to learn efficient analogue beam patterns. In our context, *multiple UEs* could cooperatively learn both a user clustering (which users share an analogue beam) and a beamforming codebook (optimal analogue weights) under a federated regime. Each agent uses its CSI, GPS and IMU data to optimize its local reward (e.g., received signal quality), and the BS aggregates these insights to update global clustering and codebook models. In this way, the MARL agents effectively solve a joint hybrid beamforming and user grouping problem, yet no raw CSI and sensor data ever leaves the devices.

4.1.4 Federated MARL for hybrid beamforming and user clustering

Hybrid beamforming divides processing between a limited number of RF chains and analogue phase shifters. Designing good analogue beam codebooks and assigning users to beams (user clustering) is crucial for maximizing throughput and mitigating interference. Conventional learning-based beamforming often assumes central CSI collection, incurring high overhead. In contrast, a federated MARL framework allows fully distributed beam training: UEs use their on-board sensors to train policies that map local state (channel and positional information) to beamforming actions, and share only aggregated model updates with the BS. Recent studies validate this approach. For example, Elbir et al. proposed FL for mmWave hybrid beamforming where users send only gradient updates of a convolutional neural network (CNN)-based beamformer to the BS, achieving robustness to channel imperfections and greatly reduced uplink overhead compared to centralized training [2]. By extending to a MARL setting, the model can also learn adaptive user grouping simultaneously, since agents implicitly coordinate through shared rewards (e.g., total throughput). The federated MARL architecture thus supports *dynamic user grouping* (or AP clustering) and *codebook learning* in one cohesive loop. Importantly, it preserves data privacy by design, because raw CSI and location remain on-device, and only model parameters are exchanged. This dramatically lowers communication overhead – unlike classical CSI feedback schemes – while still enabling the BS to learn optimized hybrid-beam strategies.

In summary, modern edge devices possess the compute and sensing capabilities to run RL locally, and federated collaborative training can then fuse these local models into a global solution. The synergy of FL and MARL is particularly powerful for wireless systems: it harnesses on-device intelligence for beamforming and clustering, while ensuring confidentiality and efficiency. As recent works note, such federated edge-artificial intelligence (AI) approaches can create a robust, versatile 6G network that meets IoT demands. In this chapter, we develop a federated MARL framework tailored to hybrid beamforming and dynamic user grouping, demonstrating that UEs can jointly train codebook and clustering algorithms at the BS – without ever sharing raw data – yielding near-optimal beam patterns with minimal overhead.

The formation of highly synchronized, low latency and spectrally efficient networks as envisioned for 6G wireless communication is critically dependent on millimetric wave (mmWave), massive MIMO, scattered intelligence (SI), and edge computing elements [3][4]. As mmWaves are degraded by atmospheric attenuation and physical obstructions, AI aided signal processing, beamforming, and adaptive antenna techniques are critical for better link reliability.

Most commonly mmWave communications utilize hybrid beamforming, where analogue beamforming manages clusters of users with similar channel characteristics, forming a single beam for each cluster. Within these clusters, users are spatially multiplexed via digital beamforming, allowing hybrid beamforming to combine the benefits of both analogue and digital beamforming while requiring fewer RF chains. This approach supports the implementation of massive MIMO, which enables the formation of narrow beams. However, these narrow beams increase beam training time and necessitate line-of-sight communication, limiting their application in scenarios with high mobility.

Massive MIMO also enables the creation of arbitrarily shaped beams, which is achieved by designing the phase values of the analogue phase shifters. This design process involves selecting an optimal set of phase shift values, referred to as a *codebook*, using appropriate algorithms. As the number of antennas in massive MIMO systems increases, the search space for phase shift vectors grows, making exhaustive search algorithms impractical. Various learning-based methods have been proposed for optimal codebook design to achieve these arbitrary beam shapes. MARL has demonstrated robustness and situational awareness in adaptive codebook design.

As already mentioned, mmWave communications use hybrid beamforming, where analogue beamforming produces a single beam for clusters of users with identical channel characteristics. By spatially multiplexing users within these clusters, digital beamforming allows hybrid beamforming to combine analogue and digital benefits with fewer RF links. It supports massive MIMO, which narrows beams. Line-of-sight communication and extended beam training limit narrow beam utilization in high-mobility environments.

Designing analogue phase shifter phase values lets huge MIMO produce any beam pattern [5]. This design method chooses the best phase shift *codebook*, or set of values, using algorithms. In large MIMO systems with multiple antennas, phase shift vector search space increases, making exhaustive search unfeasible [6]. Multi-learning-based solutions have been devised for optimal codebook development to acquire any beam morphology. MARL is robust and situational in adaptive codebook design.

4.2 Common Approaches of Codebook Design for Hybrid Beamforming in mmWave Massive MIMO Networks

Generally speaking, codebook design approaches necessitate an initial clustering phase, which frequently depends on offline data collecting and training [5] or inefficient beam training methods. When dealing with situations that are dynamic, it is vital to re-cluster frequently. It is possible for downlink codebook design algorithms to identify the ideal beam for beamforming by utilizing only uplink CSI and prior codebook entries [6]. This eliminates the need for consumers to provide any personal information. In light of this, there is a requirement for improved clustering algorithms that are capable of functioning without the direct transfer of user data to the base station. The hybrid beamforming technique is utilized in mmWave communications. In this technique, analogue beamforming is utilized to create a single beam for groups of users who share the same channel characteristics. In these clusters, users are spatially multiplexed through the use of digital beamforming, which enables hybrid beamforming to combine the advantages of analogue and digital technologies while reducing the number of RF chains. Through the use of this technology, narrow beams are produced, which permits massive MIMO. Narrow beams necessitate communication in a line of sight and longer training sessions, which limits their application in environments with high levels of mobility.

The ability of massive MIMO to generate beams of any shape is made possible by the design of analogue phase shifter phase values. For the purpose of selecting an optimal phase shift *codebook*, or set of values, this design technique requires the utilization of algorithms. It is impossible to use exhaustive search procedures in massive MIMO systems that have a large number of antennas because the phase shift vector search space increases. The generation of ideal codebooks for the purpose of obtaining arbitrary beam morphologies has been accomplished through the development of many learning-based methodologies. Multi-agent reinforcement learning, also known as MRL, is a robust and situational technique that is used in adaptive codebook design.

4.3 FL Assisted Codebook Design for Hybrid Beamforming

As reported by [7] [8], FL provides a solution by transferring the model to the data rather than the other way around, which helps to maintain the confidentiality of the account of the user. This makes it possible to take advantage of personal data, such as coarse GPS location, accelerometer, and gyroscope data, in order to enhance beamforming effectiveness while simultaneously protecting the privacy of the user [9] [10]. The purpose of this work is to improve the design of MARL-based codebooks by including FL at the user equipment level. This will make it possible to cluster users in a manner that is both more accurate and more customizable. We improve the overall performance of codebook design by training a model at each user device to dynamically assign the user to a suitable cluster. This is accomplished by utilizing data from GPS, acceleration, and orientation for the training of the model. Figure 4.1 shows a set-up of a horizontal FL server at the base station for identifying clusters at each user devices. It has two components – one a model update and the other a global update. The model update is derived from the aggregate of learning that is recorded at the base station with inputs received from all users. The CSI feed coming from different users help the system to estimate the present state of the channel and share the aggregate update of the model with all the users. Along with CSI, data from a IMU and GPS unit are used to drive the learning of the system.

Figure 4.1 illustrates the federated analogue-beam training architecture. Each user device (UE) collects private sensor data (GPS coordinates and inertial measurements) and trains a local model on these features to predict its analogue-beam cluster. The user performs local training and sends only model updates to the base station, where they are aggregated into a global model. As a result, the base station obtains a trained clustering model without ever seeing raw sensor data. This ensures that confidential sensor readings remain on-device, and only aggregate model parameters are exchanged.

Figure 4.1 Horizontal federated learning server at the base station for identifying clusters at each user device.

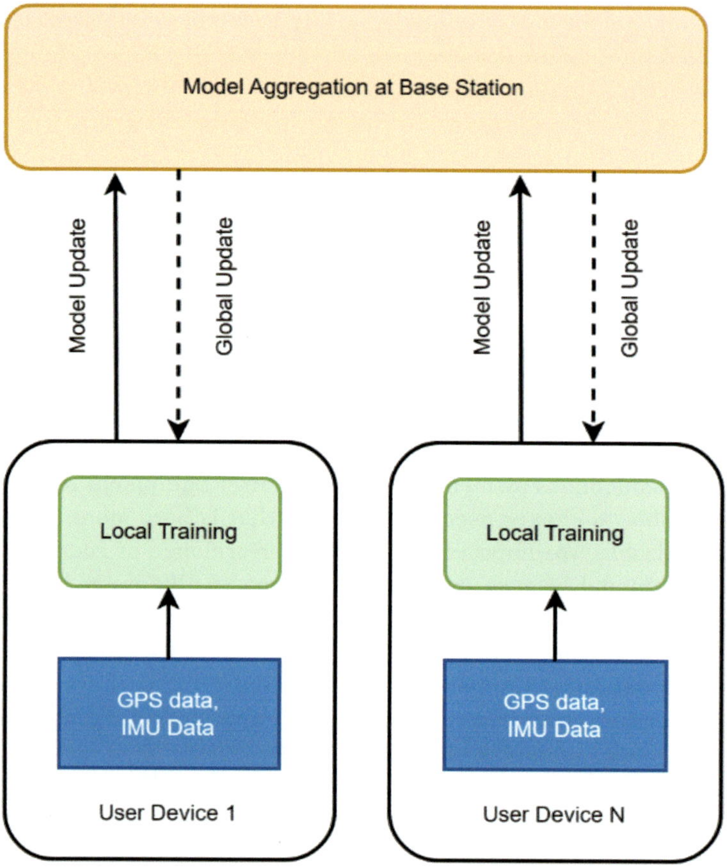

Figure 4.2 shows the MARL based beam codebook learning that takes place at the base station. Here the uplink CSI, reference signal received power (RSRP) (provides the measure of the wireless link in 4G/5G networks) feedback, automatic repeat request (ARQ) status, last used phase vector and identified cluster at user equipment (UE) level using FL are essential elements

for creation of the model aggregation. Further, there is agent level collaboration and cooperation for inter-beam interference reduction. A bi-directional link between the MARL system and the participating nodes shares the aggregated FL model, uplink channel sounding, and global update of the learning, ARQ status, RSRP feedback and identified cluster feedback for ascertaining the quality of the process. Identified cluster ID at each UE is fed back to the base station for improved MARL performance. Several such clusters are considered for establishing the effectiveness of the FL based approach.

Figure 4.2 Full system model with MARL at the base station improved by user clustering at the UE.

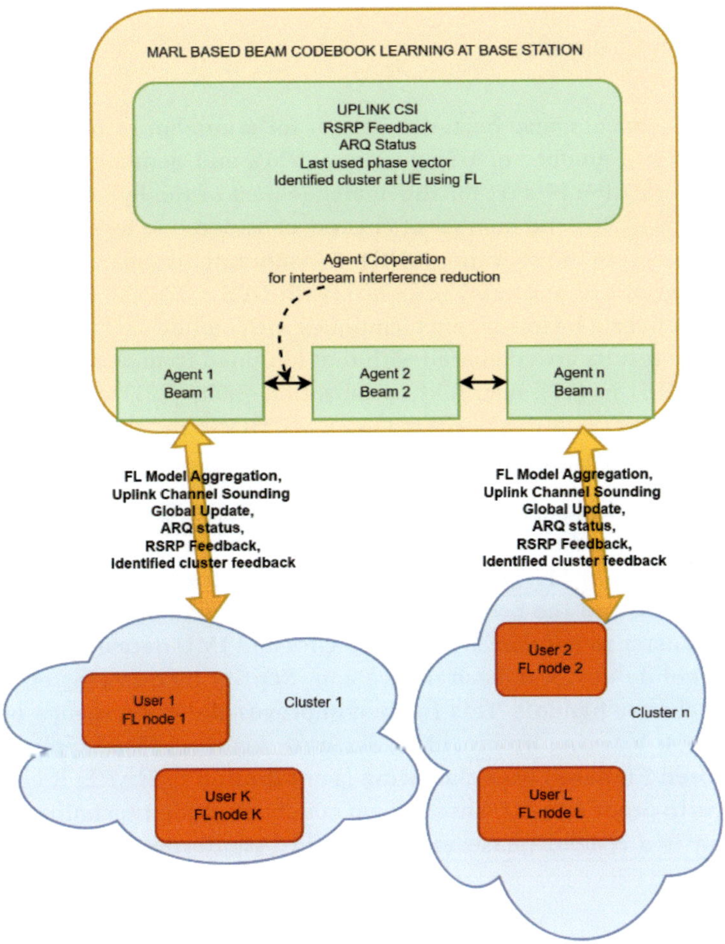

Figure 4.2 shows the MARL-based digital-beamforming stage at the base station. The base station runs multiple reinforcement-learning agents (one per beam) that jointly design a digital beam codebook based on aggregated state information. Each agent's state includes uplink feedback (CSI, RSRP), ARQ status, the previously used analogue phase vectors, and the identified cluster index for each UE (as output by the FL model). The agents cooperate to learn non-interfering beam patterns, adjusting codebook entries iteratively to improve coverage and reduce inter-beam interference. This cooperative MARL procedure yields an environment-specific set of digital and analogue beam codebooks that complement the privacy-preserving analogue clusters.

4.4 Experimental Results

Certain experiments have been carried out for a number of base station (BS) antennas (NBS), number of UE antennas (NM), and number of UEs (M) for certain quantization bits (r) for the analogue part of the hybrid beamforming with numbers of BSs and number of clusters considered to be equal. Figure 4.3 shows a signal to noise ratio (SNR) vs. sum-rate comparison with noisy measurement of CSI and array response (with NBS = 32, NM = 4, r = 3, M = 4) for different hybrid beamforming techniques with mobile user moving between clusters. The results are compared with that obtained from deep reinforcement learning (DRL) in [11], manifold optimization (MO) [12], sparse orthogonal matching pursuit (SOMP) algorithm [13], CNN-MIMO and the two-stage hybrid beamforming (TS-HB) algorithm [14].

The results indicate that for mobile users, FL-based user clustering with MARL outperforms traditional initial access-based beam clustering in MARL, as shown in [11]. This is largely due to the slower nature of traditional beam sweeping and the non-instantaneous cluster handoff when users move from one cluster to another. With coarse GPS and IMU data available at edge devices (mobile devices), each device can identify its own cluster, enabling seamless cluster handoff. This further improves the convergence of MARL for codebook learning by providing accurate feedback. The applicability of the proposed FL-based user clustering is not limited to MARL; it can also be utilized with other non-ML-based beam codebook design techniques, as user clustering is a crucial phase in any hybrid beamforming codebook design approach.

Figure 4.3 SNR vs. sum-rate comparison with noisy measurement of CSI and array response (NBS = 32, NM = 4, r = 3, M = 4) for different hybrid beamforming techniques with mobile user moving between clusters.

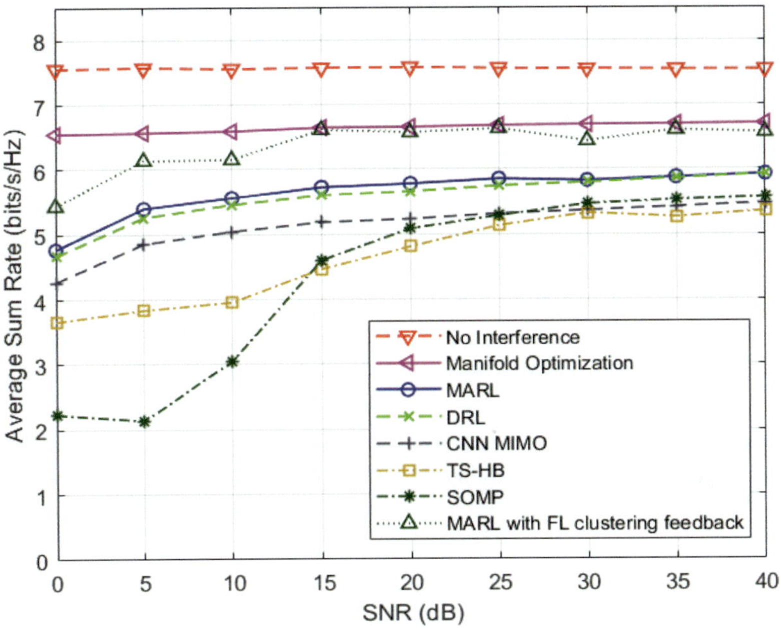

4.5 Conclusion

In this chapter, we have discussed a FL driven MARL approach for edge-intelligent beamforming codebook design for mmWave massive MIMO setups which is very crucial for next-generation wireless systems. The proposed federated MARL architecture facilitates the integrated optimization of dynamic user grouping (or access point clustering) and codebook learning within a unified framework. Crucially, it ensures inherent data privacy by retaining raw CSI and location information locally at user devices, transmitting only model parameters to the central server. This approach substantially reduces communication overhead compared to traditional CSI feedback mechanisms while still enabling the BS to effectively learn optimized hybrid beamforming strategies. The simulation results demonstrate that, for mobile users, FL-based user clustering integrated with MARL yields superior performance compared to conventional initial access-based beam clustering approaches within MARL frameworks.

References

[1] B. Banerjee, R. C. Elliott, W. A. Krzymień and M. Medra, "Access Point Clustering in Cell-Free Massive MIMO Using Conventional and Federated Multi-Agent Reinforcement Learning," in IEEE Transactions on Machine Learning in Communications and Networking, vol. 1, pp. 107–123, 2023, doi: 10.1109/TMLCN.2023.3283228.

[2] A. M. Elbir and S. Coleri, "Federated Learning for Channel Estimation in Conventional and RIS-Assisted Massive MIMO," in IEEE Transactions on Wireless Communications, vol. 21, no. 6, pp. 4255–4268, June 2022, doi: 10.1109/TWC.2021.3128392.

[3] Wang, C.X., You, X., Gao, X., Zhu, X., Li, Z., Zhang, C., Wang, H., Huang, Y., Chen, Y., Haas, H., et al. On the Road to 6G: Visions, Requirements, Key Technologies, and Testbeds. IEEE Commun. Surv. Tutorials 2023, 25, 905–974, doi:10.1109/COMST.2023.3249835.

[4] You, X., Wang, C.X., Huang, J., Gao, X., Zhang, Z., Wang, M., Huang, Y., Zhang, C., Jiang, Y., Wang, J., et al. Towards 6G Wireless Communication Networks: Vision, Enabling Technologies, and New Paradigm Shifts. Sci. China Inf. Sci. 2021, 64, 1–74, doi:10.1007/s11432-020-2955-6.

[5] Zhang X., et. al, "Deep Reinforcement Learning for Over-the-Air Federated Learning in SWIPT-Enabled IoT Networks," 2022 IEEE 96th Vehicular Technology Conference (VTC2022-Fall), London, United Kingdom, 2022, pp. 1–5, doi: 10.1109/VTC2022-Fall57202.2022.10012702.

[6] Adhikary, A., et al. "Joint spatial division and multiplexing for mm-wave channels." IEEE Journal on Selected Areas in Communications 32.6 (2014): 1239–1255.

[7] Yue Z., et al. "Federated Learning with Non-IID Data." ArXiv, abs/1806.00582 (2018). https://doi.org/10.48550/arXiv.1806.00582.

[8] Tian Li et al. "Federated Learning: Challenges, Methods, and Future Directions." IEEE Signal Processing Magazine, 37 (2019): 50–60. https://doi.org/10.1109/MSP.2020.2975749.

[9] McMahan, B., et al. "Communication-efficient learning of deep networks from decentralized data." Artificial intelligence and statistics. PMLR, 2017.

[10] Truong, Nguyen, et al. "Privacy preservation in federated learning: An insightful survey from the GDPR perspective." Computers & Security 110 (2021): 102402.

[11] Zhang, Yet. Al., A. Reinforcement Learning of Beam Codebooks in Millimeter Wave and Terahertz MIMO Systems. IEEE Transactions on Communications. 2022, 70, 904–919.

[12] Yu, X et. al.. Alternating Minimization Algorithms for Hybrid Precoding in Millimeter Wave MIMO Systems. IEEE Journal of Selected Topics in Signal Processing. 2016, 10, 485–500.

[13] Ayach, O.E.; Rajagopal, S.; Abu-Surra, S.; Pi, Z.; Heath, R.W. Spatially Sparse Precoding in Millimeter Wave MIMO Systems. IEEE Transactions on Wireless Communications. 2014, 13, 1499–1513.

[14] Alkhateeb, A.; Leus, G.; Heath, R.W. Limited Feedback Hybrid Precoding for Multi-User Millimeter Wave Systems. IEEE Transactions on Wireless Communications. 2015, 14, 6481–6494.

CHAPTER

5

Advancements in Cardiac Monitoring: The Role of AI and 5G Networks in Optimizing Cardiovascular Healthcare

Amrita Choudhury*, Kandarpa Kumar Sarma and Lachit Dutta

Department of Electronics and Communication Engineering, Gauhati University, Guwahati-781014, Assam, India
* Corresponding author: ch.amrita773@gmail.com

Email: kandarpaks@gauhati.ac.in; lachit@gauahti.ac.in

Abstract

Cardiovascular diseases have remained one of the major health burdens worldwide and their timely detection and thus accurate diagnosis have become key elements in optimizing patient outcomes. In this chapter, we review artificial intelligence applied with 5G networks for cardiac monitoring. While artificial intelligence and data analytics have improved diagnostic accuracy and treatment planning for invasive techniques, non-invasive approaches, such as wearable devices and remote monitoring, have been increasingly exhibiting accuracy in the early identification of potential problems. However, cardiac monitoring with the use of AI and 5G technology is yet to overcome many challenges. These include a host of issues concerning high-quality data acquisition in large quantities, much-needed standardization and interoperability among disparate devices and platforms, privacy and security

DOI: 10.1201/9788743809197-5

issues, and finally, the cost-effectiveness of the proposed system. Further research and development on AI and 5G technology will likely bring enormous change to coronary healthcare and results for patients.

5.1 Introduction

Cardiovascular diseases remain one of the leading causes of death globally, highlighting the urgent need for early detection and accurate diagnosis to improve patient outcomes. In this context, artificial intelligence (AI) has emerged as a transformative tool in cardiac healthcare, offering the capability to deliver precise diagnoses, predict disease progression, and enable personalized treatment. By analysing vast quantities of patient data, including medical history, laboratory results, and imaging data, AI algorithms can detect complex patterns and generate predictions that empower healthcare professionals to make more informed decisions. Consequently, AI-assisted diagnostic support holds immense potential to revolutionize the diagnosis, treatment, and management of cardiovascular diseases.

A critical enabler of these AI-driven advancements is the rapid evolution of communication technologies, particularly fifth-generation (5G) networks. With their high bandwidth, low latency, and capacity for massive connectivity, 5G networks enable fast and secure data exchange between hospitals, diagnostic labs, and remote patient-monitoring devices. This robust connectivity is crucial for managing large cardiac imaging datasets and conducting real-time analyses of vital signs collected from wearable sensors. Furthermore, the low-latency environment provided by 5G networks supports near-instantaneous feedback loops between AI algorithms and clinicians, facilitating timely and potentially lifesaving interventions in critical cardiac care scenarios.

The application of AI-aided diagnostics in cardiac healthcare has already shown remarkable potential, particularly in a field where even small errors can have severe consequences for patients. AI systems are capable of predicting the likelihood of heart disease, stroke and other cardiovascular conditions, optimising treatment plans, and forecasting patient outcomes. As these technologies continue to advance, the development of more sophisticated algorithms and tools is anticipated to further enhance healthcare providers' ability to improve patient outcomes. However, the successful deployment of such systems hinges on the availability of high-speed, reliable data transfer across both central facilities and remote or underserved areas, underscoring the importance of robust connectivity solutions like 5G networks.

Despite the enormous promise of AI-aided diagnostics in cardiac healthcare, several challenges must be addressed to fully realize its benefits. One critical issue is the need for high-quality data to train AI algorithms effectively [1]. Another significant challenge lies in algorithmic bias, which can lead to inaccurate diagnoses and suboptimal treatment recommendations. Ethical concerns also come to the forefront, particularly regarding patient data privacy and the fear that AI could replace human healthcare professionals [2]. This chapter delves into the current state of AI-assisted cardiac diagnostics, exploring its diverse applications and the challenges that must be overcome to unlock its full potential. Furthermore, it highlights how robust connectivity solutions, such as 5G networks, are pivotal in supporting these advancements.

5.2 Related Background

AI-aided diagnostic support in cardiac healthcare aims to enhance patient outcomes by leveraging AI algorithms to analyse vast amounts of patient data, enabling more accurate, personalised diagnoses, treatment plans, and disease management [3]. This area of research focuses on harnessing AI to gain deeper insights into patient health, empowering healthcare providers to make better-informed decisions and ultimately improving patient outcomes.

AI-assisted diagnostics for heart health are underpinned by significant advancements in machine learning (ML) [4], big data analytics [6], and healthcare technology. These innovations have facilitated the creation of powerful algorithms capable of processing extensive patient datasets and identifying patterns that may be missed by human clinicians. Deep learning (DL) [5], a subset of ML, leverages artificial neural networks to derive insights from large datasets and is a fundamental driver of AI applications in diagnostics. DL has proven particularly effective in analysing medical imaging data, such as CT scans and MRIs, to identify diseases and abnormalities. Similarly, big data analytics [6] has been instrumental in enabling AI-assisted diagnostics. With the widespread digitisation of healthcare data, enormous volumes of patient information are now available for analysis. Big data tools empower healthcare professionals to uncover intricate patterns and relationships in this data that traditional statistical methods might fail to detect.

In parallel, the emergence of 5G networks has provided significant support for these AI-driven healthcare innovations. By enabling high-speed data transfer and low-latency communication, 5G networks facilitate the real-time sharing and analysis of large medical imaging files, continuous patient-monitoring data, and other critical information. This capability is particularly valuable in urgent or resource-limited scenarios, supporting remote consultations, telemedicine services, and AI-based cardiac diagnostics.

Through 5G-enabled infrastructures, clinicians and AI systems can exchange insights quickly and efficiently, enhancing the timeliness and precision of medical decision-making. This robust connectivity also enables seamless integration of AI tools into the healthcare workflow, ensuring that innovations in AI and 5G technologies contribute directly to improving patient care.

5.2.1 Basic principle of cardiac monitoring

Cardiac monitoring is a medical procedure focused on tracking and recording the heart's electrical activity to diagnose and manage heart-related conditions. It can be categorised into two primary types: clinical-based monitoring and data-processing-based monitoring.

5.2.1.1 Clinical based monitoring

Clinical-based cardiac monitoring involves tracking and recording the heart's electrical activity using specialised instruments such as electrocardiography (ECG) machines or Holter monitors [1]. This method detects the electrical impulses generated by the heart and creates graphs, known as ECGs, which healthcare professionals analyse to diagnose cardiac conditions and develop personalised treatment strategies. During an ECG, small electrodes are placed on the chest, arms, and legs, adhering to the skin. These electrodes are connected to an ECG machine, which records the heart's electrical activity as it beats. The machine produces a graph of electrical activity that healthcare providers examine to identify irregularities in the heart's rhythm or other signs of heart-related illnesses. Holter monitors, another clinical cardiac monitoring device, are designed to continuously record the heart's electrical activity over an extended period, typically 24 to 48 hours [7]. This portable device includes electrodes attached to the chest and is worn by the patient throughout the monitoring period. The monitor continuously captures the heart's electrical activity, and the recorded data is later analysed by healthcare providers to detect rhythm abnormalities or other indicators of heart-related disorders. Clinical-based cardiac monitoring provides essential insights into the heart's function, enabling healthcare providers to diagnose heart diseases accurately and create customised treatment plans. This approach is especially useful for diagnosing intermittent conditions or evaluating the effectiveness of treatments over time.

5.2.1.2 Data processing based monitoring

Data processing-based cardiac monitoring leverages AI algorithms to analyse vast quantities of patient data, delivering more precise and individualised diagnoses and treatments [8]. This approach uses ML techniques to identify patterns and anomalies in patient data—such as ECG readings, vital signs, and medical histories—to detect heart-related conditions and enable timely interventions. AI algorithms are trained on extensive datasets of patients with known heart-related conditions, learning to recognise patterns and correlations associated with these conditions. Once trained, the algorithms can analyse new patient data and provide real-time insights to healthcare providers, facilitating informed decision-making and personalised treatment plans [8]. This method enhances the ability of healthcare providers to detect heart-related conditions early, often before symptoms become life-threatening. By providing accurate diagnoses and tailored treatment strategies, AI-powered monitoring can also reduce unnecessary medical tests, procedures, and hospitalisations, minimising the need for extensive manual data analysis. Data-processing-based cardiac monitoring holds significant potential for improving patient outcomes by enabling earlier and more accurate detection and management of heart-related issues. However, to maximise its effectiveness, challenges such as ensuring high-quality data for training and addressing potential biases in AI algorithms must be resolved.

5.2.2 Different methods of cardiac monitoring

Cardiac monitoring methods can be broadly divided into two categories: invasive and non-invasive techniques [9].

5.2.2.1 Invasive methods

Invasive methods of cardiac monitoring involve inserting a catheter into the heart through a blood vessel. Imaging techniques, such as X-ray or echocardiography, are used to guide the catheter to the heart. Once positioned, the catheter measures various cardiac parameters, including blood pressure, oxygen levels, and electrical activity [1]. Here are some common examples of invasive methods of cardiac monitoring:

- **Cardiac catheterisation:** This procedure involves inserting a catheter through a blood vessel in the arm or leg, which is then guided to the heart. Once correctly positioned, the catheter allows for the measurement of blood pressure and oxygen levels in different areas of the heart and facilitates interventions such as angioplasty or stent placement.

- **Swan–Ganz catheterisation:** This type of cardiac catheterisation involves inserting a catheter into the pulmonary artery. It measures pressures in the heart and lungs, aiding in the diagnosis and management of conditions like heart failure and pulmonary hypertension [10].
- **Electrophysiology study:** This procedure involves guiding a catheter to the heart to map its electrical activity. It helps diagnose and treat arrhythmias by pinpointing the source of abnormal electrical signals [11].
- **Myocardial biopsy:** This procedure involves removing a small piece of heart tissue for examination. It is typically performed using a catheter inserted through a blood vessel and is used to diagnose conditions like myocarditis [11].
- **Coronary angiography:** This invasive cardiac monitoring method uses X-ray imaging and a contrast agent to visualise the heart's blood vessels. The procedure begins with administering a mild sedative and local anaesthesia, followed by inserting a catheter into an artery. A contrast agent is injected to highlight blockages, and if a blockage is detected, balloon angioplasty may be performed, possibly accompanied by stent placement. After the procedure, the catheter is removed, and pressure is applied to the insertion site to prevent bleeding [10].

Invasive methods of cardiac monitoring are typically reserved for serious or complex conditions and carry potential risks, including bleeding or infection. Despite these risks, they offer more detailed information about the heart compared to non-invasive methods and enable interventions such as angioplasty or stent placement.

5.2.2.2 Non-invasive method

Non-invasive cardiac monitoring involves using external sensors or equipment to measure various heart parameters. While these methods may not offer the same level of detail as invasive techniques, they are generally safer and more convenient [9]. Here are some examples of non-invasive cardiac monitoring methods:

- **Electrocardiography:** An ECG is a non-invasive procedure that painlessly records the heart's electrical activity by placing electrodes on the skin. It is used to identify and treat various heart conditions, including arrhythmias, heart attacks, and heart failure [1, 2].
- **Holter monitoring:** Holter monitoring is a type of ECG that involves carrying a portable device to continuously monitor the heart's electrical activity for 24 to 48 hours. It is used to diagnose and treat heart disorders that may not be detectable during a standard ECG [1].
- **Event monitoring:** Event monitoring is a type of ECG that involves carrying portable equipment to record the heart's electrical activity over an extended period, such as 30 days or more. It is used to diagnose and manage rare conditions, such as cardiac arrhythmias and palpitations.
- **Echocardiography:** Echocardiography is a non-invasive test that uses ultrasonic waves to create images of the heart. It is used to detect and manage conditions such as heart valve problems, heart failure, and congenital heart defects.

- **Stress testing:** Stress testing involves exercising on a treadmill or stationary bike while monitoring the heart's electrical activity, blood pressure, and oxygen levels. It is used to diagnose and manage conditions such as coronary artery disease and heart failure.

Non-invasive cardiac monitoring methods are generally painless and safe, avoiding the risks associated with invasive procedures (Table 5.1). The choice of monitoring technique depends on the patient's specific condition and the information needed for diagnosis or treatment.

Table 5.1 The classification of invasive and non-invasive methods of cardiac monitoring:

Ref	Methods of cardiac monitoring	Invasive methods	Non-invasive methods
[1]	ECG		✓
[1]	Holter monitoring		✓
[12]	Event monitoring		✓
[12]	Echocardiography		✓
[12]	Stress testing		✓
[10]	Cardiac catheterization	✓	
[10]	Coronary angiography	✓	
[11]	Electrophysiology	✓	
[10]	Intravascular ultrasound (IVUS)	✓	
[12]	Magnetic resonance imaging (MRI)		✓
[12]	Computed tomography angiography		✓
[11]	Pacemaker	✓	
[13]	Intracardiac echocardiography	✓	
[13]	Transoesophageal echocardiography	✓	
[13]	Pulmonary artery catheterization	✓	

5.3 Current Practices of Adoption of AI Tools with Cardiac Monitoring

Recent advancements in applying AI to both invasive and non-invasive cardiac monitoring have significantly improved healthcare standards. This section explores recent studies and innovations in the field, highlighting how AI has effectively enhanced diagnostic accuracy, treatment planning, and patient outcomes in cardiac healthcare.

5.3.1 Invasive cardiac monitoring method

Invasive cardiac monitoring methods, such as coronary angiography, can greatly benefit from AI integration. AI algorithms can automatically analyse images obtained during procedures to detect abnormalities or blockages in the heart's blood vessels with high accuracy. AI-powered catheters are also being utilised in catheter ablation procedures to detect and diagnose abnormal heart rhythms in real time. Additionally, AI can analyse data from intracardiac electrograms (ICEGs) acquired during electrophysiology procedures to enhance diagnostic precision. Furthermore, AI has been employed to predict outcomes of invasive cardiac procedures, such as assessing the risk of complications following percutaneous coronary intervention (PCI), enabling more informed clinical decision-making.

The article [14] highlights the promising applications of AI in interventional cardiology, showcasing how AI technologies are advancing diagnosis, treatment planning, and procedural outcomes in this field. AI has already found utility in robot-assisted percutaneous coronary intervention and the evaluation of coronary artery ischemia and plaque burden through intravascular ultrasonography. While DL systems have been created for cardiovascular purposes, their influence on cardiology practices remains uncertain. They still face limitations, including challenges in generalizability and decision-making pertaining to heart anatomy. The authors of [15] showed that ML can help anaesthesiologists perform transoesophageal echocardiography (TEE). AI in point of care ultrasounds (POCUS), like TEE, improves accuracy and efficiency, bridges the gap between technological capabilities, and provides a reliable ultrasound-based diagnosis. POCUS is becoming more significant in resident physician training for intraoperative and perioperative cardiac and pulmonary function assessment. AI integration promises to improve patient outcomes by allowing doctors to make faster evaluations utilising more data and receive diagnostic and treatment recommendations. In [16], the authors present a novel automated system for detecting misplaced medical tubes in chest radiography (CXR) using DL techniques. Their proposed method utilizes an efficient convolutional neural network (CNN) architecture called EfficientNet. Through meticulous experimentation, the system achieves an impressive average accuracy of 0.95 in accurately detecting and classifying improperly positioned tubes on CXR images. The content of [17] is related to the importance of early detection of malposition catheters and lines in hospital patients, especially in the current COVID-19 pandemic. The authors present an analysis of their solution to the "RANZCR CLiP: Catheter and Line Position Challenge" on Kaggle, achieving a high accuracy score of 0.972. Their solution is based on a transfer learning-based CNN that is smaller in size and faster than other architectures, heavily inspired by UNet and EfficientNet. The study [18] reports the design of an artificial neural network (ANN) to perform automated pressure waveform analysis for real-time identification of damping during coronary angiography. It has been found to be highly accurate in classifying beats, with 99.4% accuracy when compared with an internal core laboratory and 98.7% accuracy when compared with an external core laboratory. This study shows that ML can assist with patient safety and quality control of procedures. In the study [19], a novel platform named AtheroEdge (AE3.0DL) is introduced as a windows-based system that employs multiclass DL methods. The study has collected data from 500 patients, including carotid ultrasound and corresponding coronary angiography scores. Thirty-nine covariates are grouped into three clusters: office-based, laboratory-based, and carotid ultrasound image phenotypes. The results show impressive accuracy rates of 95.0% and 95.34% for the respective clusters, marking a significant improvement of 32.93% and 9.94% compared to CCVRC and ML methods, respectively. Remarkably, AE3.0DL operates within just 1 second. In [20], the authors introduce and evaluate a novel explainable DL model called CAD-DL, specifically designed to detect obstructive coronary artery disease (CAD) using single-photon emission computed tomography (SPECT) myocardial perfusion imaging (MPI). The study demonstrates that the CAD-DL model outperforms standard quantitative analysis and clinical visual reading methods in diagnostic accuracy for MPI. Furthermore, the model can be easily integrated into conventional clinical software and delivers rapid results, making it highly suitable for routine clinical workflows.

The work [21] explores the utilisation of AI in cardiology, particularly in cardiac electrophysiology (EP), focusing on automatic arrhythmia detection, prognosis, management, and the identification of new EP concepts. Topics include arrhythmia and disease classification based on electrocardiograms, detection of atrial fibrillation sources, substrate assessment, risk evaluation for atrial fibrillation and ventricular tachyarrhythmias, and predicting outcomes of cardiac resynchronization therapy. The authors in [22] have proposed a tracked guidewire or catheter-based ultrasound-guided vascular navigation system. To address this challenge, a DL-based has been approached to segment the vessel lumen from ultrasound images has been adopted. After post-processing, a 90% accuracy rate has been achieved, and the system is capable of being integrated into an image-guided surgical navigation mechanism. The study [23] presents a DL model, called RHCNet, that detects elevated mean pulmonary capillary wedge pressure (mPCWP) using only the standard 12-lead ECG. The model demonstrated strong discriminatory ability; however, its performance varies depending on the specific indication.

5.3.2 Non-invasive cardiac monitoring method

Non-invasive methods of cardiac monitoring include electrocardiography (ECG), echocardiography, cardiac MRI, and cardiac CT. AI techniques have been increasingly applied to these methods to enhance their accuracy and efficiency. Recent studies highlight promising advancements, such as the use of DL algorithms for analysing ECG data and ML algorithms for processing echocardiography images. These innovations improve diagnostic precision and streamline workflows. Overall, integrating AI techniques into non-invasive cardiac monitoring holds significant potential to enhance the accuracy and efficiency of diagnosing and treating cardiac diseases.

The study [24] introduces DEEPVESSEL-FFR, a platform utilising a DL technique to calculate fractional flow reserve (FFR) values from computed tomography angiography (CTA) images, demonstrating higher diagnostic performance compared to wire-based FFR. Secondary evaluations indicate high accuracy, sensitivity, specificity, negative predictive value, and moderate positive predictive value. In [25], researchers developed and validated a DL system for quantifying atherosclerotic plaque using coronary CT angiography (CCTA). The system accurately measured plaque volume and stenosis severity, showing close agreement with expert readings and intravascular ultrasound assessments. It also demonstrated potential prognostic value in predicting future myocardial infarctions, highlighting its utility in cardiovascular risk assessment. The study [26] evaluated the use of DL image reconstruction (DLIR) in reducing radiation doses during CCTA. Results showed that DLIR effectively lowered radiation exposure without increasing image noise while maintaining high reliability in stenosis severity and plaque composition analysis. In [27], a single algorithm combining DL and ML was developed to predict in-hospital mortality in Asian patients with acute coronary syndrome (ACS), using predictors such as age, heart rate, Killip class, cardiac catheterisation, oral hypoglycaemic agents, and antiarrhythmic agents. The authors of [28] demonstrated that ML analysis of three-dimensional cardiac motion can predict survival and mechanisms of right ventricular failure in pulmonary hypertension, independent of conventional risk factors. This approach aids clinicians in stratifying patients and identifying those at higher risk of adverse outcomes. In [29], a novel method for detecting atrial fibrillation (AFib) was proposed using a 2D image representation of the 1D ECG signal, pre-processed and classified via a bi-directional long short-term memory (LSTM) network. The method achieved an accuracy of 98.85% and can detect AFib using ECG segments as short as 4 seconds, supporting reliable and timely automated detection. In [30], researchers evaluated an AI model's effectiveness in detecting hidden atrial fibrillation (AF) using 24-hour Holter monitoring data during normal sinus rhythm. The AI model outperformed conventional methods of evaluating supraventricular ectopy (SVE) burden, particularly during nighttime recordings, highlighting its potential for identifying occult AF episodes. The study [31] assessed EchoNet-Dynamic, a DL model for analysing echocardiography images, focusing on left ventricle (LV) segmentation and ejection fraction (EF) prediction from point-of-care ultrasound (POCUS) apical four-chamber images. Performance deterioration was noted with images from resuscitation cases, suggesting the need for training models on datasets specific to POCUS. The study [32] introduced an AI system for automating stress echocardiography analysis, extracting novel geometric and kinematic features to train an ensemble ML classifier. The system achieved satisfactory accuracy and improved inter-reader agreement, confidence, and sensitivity by 10% in detecting severe coronary artery disease. Lastly, researchers in [33] developed algorithmic models using DL techniques like MobileNetV2 and VGG16 to analyse ECG tracings and predict cardiovascular diseases. After parameter optimisation, both algorithms achieved similar validation accuracy, and performance remained excellent even when implemented on a Raspberry Pi, albeit with slight accuracy reductions.

5.4 Real-time Cardiac Monitoring System Driven by 5G

Figure 5.1 A schematic diagram of a real time cardiac monitoring system driven by 5G.

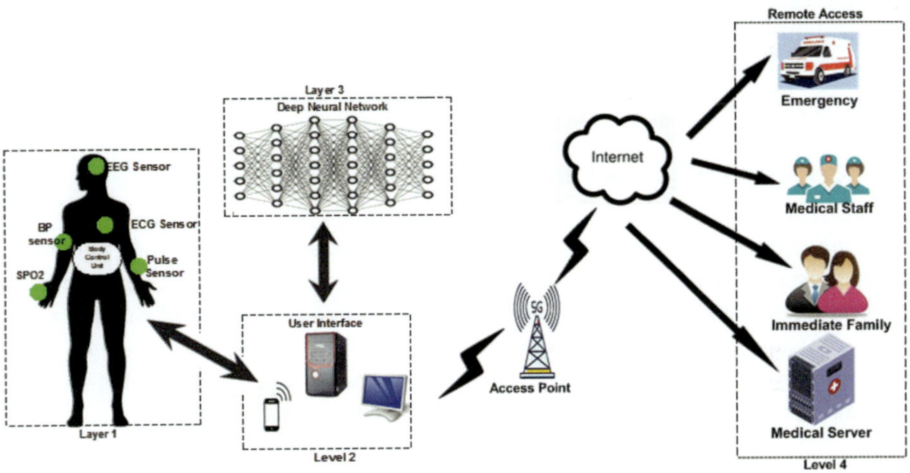

The real-time cardiac monitoring system driven by the 5G network is an advanced medical device designed to monitor the cardiac activity of patients in real-time. Powered by the latest 5G wireless technology, the system ensures fast data transfer rates and dependable connectivity. The system includes several key components: sensors, a body control unit (BCU), and a user interface. The sensors are placed on different parts of the patient's body to detect and record physiological signals. These signals are then transmitted wirelessly to the BCU, a small and portable device that patients can easily carry around. The BCU analyses the data and generates alerts if necessary, enabling continuous monitoring of the patient's cardiac activity. A user interface is also part of the system, providing medical professionals access to the data collected by the sensors and the BCU. This interface can be accessed through a web-based platform or a mobile application, allowing real-time information to be viewed from anywhere in the world. The user interface displays data such as heart rate, rhythm, and any detected abnormalities in the patient's cardiac activity. At the core of the system is a DL algorithm designed to process the large amounts of data collected by the sensors. This algorithm is trained on extensive datasets of cardiac activity to detect patterns and anomalies with high accuracy, enabling it to diagnose potential abnormalities in the patient's heart function effectively. Compared to conventional cardiac monitoring systems, the real-time cardiac monitoring system offers several significant advantages. It is entirely wireless, eliminating cumbersome cables and reducing the risk of infection. Additionally, its use of 5G technology ensures reliable and fast connectivity, even in areas with poor network coverage. This real-time cardiac monitoring system, powered by 5G, is a groundbreaking medical device with the potential to revolutionise cardiac care. By providing real-time monitoring and alerts, the system helps medical professionals detect and respond to cardiac abnormalities swiftly, potentially saving lives and improving patient outcomes. A schematic diagram of an AI-aided cardiac monitoring system is shown in Figure 5.1.

5.4.1 Processing unstructured data from wearable sensors through 5G networks

Dealing with unstructured data from wearable sensors over 5G networks presents a significant challenge due to the immense volume and diverse formats of data these sensors generate, coupled with the need for fast and reliable real-time network handling. Wearable sensors generate vast amounts of data in various formats, including text, audio, video, and sensor readings. Effectively managing this data involves a multi-step process:

- **Data collection:** Aggregating data from multiple sensors in real-time via 5G networks.
- **Pre-processing:** Cleaning and organising the raw data to remove redundancies, noise, and errors, ensuring the data's quality and reliability.
- **Transformation:** Converting unstructured data into a structured format suitable for analysis and processing, enabling accurate insights and informed decision-making.

This structured transformation is crucial for efficiently utilising the data in applications like real-time health monitoring and diagnostics. After pre-processing, securely and reliably transmitting the data through 5G networks becomes vital. The high-speed, low-latency capabilities of 5G networks facilitate the real-time analysis and processing of data, ensuring timely and accurate outcomes critical for healthcare applications.

Two key features of 5G, beamforming and network slicing, hold significant potential to revolutionise healthcare services [34]. Beamforming enhances the efficiency and effectiveness of 5G networks, while slicing enables flexible and efficient bandwidth allocation. Together, these properties support the fast and reliable data transmission required for real-time monitoring and diagnosis and contribute to the growing adoption of the Internet of Things (IoT) in healthcare. Despite its promise, deploying 5G networks requires substantial infrastructure investment, and access to this technology is not yet universal [35]. Additionally, processing unstructured data from wearable sensors via 5G networks demands advanced analytics tools capable of handling big data. Tools such as ML algorithms, DL neural networks, and natural language processing (NLP) techniques are critical for identifying data patterns and trends, extracting actionable insights, and making predictions to support decision-making [36]. The study [37] highlights that the low latency and high bandwidth of 5G networks, combined with DL algorithms, can assist in detecting atypical behaviours in elderly individuals. However, for these systems to be effective, they must be designed to respect and accommodate the preferences and needs of elderly users. CNNs and LSTMs are used to detect anomalous behaviour in old people, and a decision-making trial and evaluation laboratory (DEMATEL) analyses the most important criteria. The LSTM outperformed the CNN with 96% accuracy and identified depressive disorder as the main cause of aberrant behaviour in elderly people. The study [38] describes a novel technique for high-performance modelling in a student health monitoring system that employs ML analysis based on 5G networks. This system collects input data in the form of EEG brain waves, which are then processed and extracted using K-adaptive reinforcement learning. A naive Bayes gradient feed-forward neural network is then used to classify the extracted features. In terms of accuracy, precision, recall, F1 score, RMSE, and MAP, the efficacy of this system is compared to existing techniques.

5.4.2 Data acquisitions tools for cardiac monitoring

There are several data acquisition tools available for heart monitoring that are relevant for wearable devices. Table 5.2 summarises the data acquisition tools for cardiac monitoring relevant for wearable devices.

Table 5.2 Summarising the data acquisition tools for cardiac monitoring relevant for wearable devices.

Ref	Data acquisition tools	Principle of measurement	Description	Application
[2]	ECG	Electrical activity of the heart	A non-invasive test conducted by placing electrodes on the skin to measure the electrical activity of the heart.	Detect abnormal cardiac rhythms, diagnose heart disease, and assess treatment efficacy.
[39]	Photoplethysmography (PPG)	Blood volume changes in the skin	Non-invasive optical method that measures changes in skin's blood volume using light.	Measure heart rate, blood pressure, and blood oxygen saturation.
[39]	Ballistocardiograph (BCG)	Mechanical activity of the heart	Non-invasive method that uses sensors that are attached to the body to track the mechanical activity of the heart.	Detect abnormal cardiac rhythms and evaluate heart valve function.
[39]	Impedance cardiography (ICG)	Blood flow-related changes in electrical impedance	Non-invasive method for measuring changes in electrical impedance caused by blood flow in the heart.	Quantify cardiac output, stroke volume, and additional hemodynamic parameters.
[40]	Magnetic resonance imaging (MRI)	Detailed images of the heart	By utilising a powerful magnetic field and radio waves, detailed images of the heart are generated.	Diagnosis and monitoring of heart disease, and treatment effectiveness evaluation.

5.4.3 Data analytics techniques used in cardiac monitoring

Data analytics in heart monitoring involves the use of advanced computational techniques and algorithms to analyse the large amount of data generated from cardiac monitoring devices. The aim is to extract meaningful insights and patterns from the data to improve diagnosis, treatment, and disease management. Table 5.3 shows the summary of different data analytics techniques used in cardiac monitoring

Table 5.3 Summarizing data analytics techniques used in cardiac monitoring.

Ref	Techniques	Description
[41]	Machine learning	A subfield of AI involving the training of algorithms to recognise data patterns and make predictions or decisions based on this recognition can be used to diagnose cardiovascular conditions and predict outcomes.
[42]	Deep learning	A subfield within ML that entails training neural networks with multiple layers to identify patterns in data can be effectively applied for tasks such as image analysis and detecting anomalies in ECG signals.
[43]	Natural language processing	A subset of AI concerned with the analysis and comprehension of human discourse can be used for analysing clinical notes and extracting heart-related information.
[44]	Predictive modelling	The method of analysing data and making future predictions using statistical models can be used to forecast the likelihood of cardiac disease or patient outcomes.
[45]	Signal processing	Analysing and extracting relevant information from physiological signals, such as ECG or blood pressure data, using mathematical algorithms. Identifies irregularities or patterns in cardiac activity.
[42]	Time series analysis	A statistical method for analysing time-series data can be employed to examine ECG signals and identify irregularities or changes over time. This approach allows for the detection of patterns, trends, and anomalies in the ECG data, providing valuable insights into the heart's electrical activity and enabling early identification of potential cardiac issues.
[46]	Data visualization	The procedure for generating data visualisations can aid clinicians in identifying patterns or anomalies in cardiac monitoring data.

5.5 Proposed Approach: AI Aided Critical Healthcare Decision Support Using a Temporal Gated Deep Network

The implementation framework for the temporal gated deep network (TGDN) is carefully structured to manage the intricacies of healthcare data and provide dependable decision support in critical situations, as illustrated in Figure 5.2. The process begins with input processing, where raw patient data is passed through a specialised healthcare data pre-processor. This module plays a vital role in ensuring the data is cleaned, normalised, and encoded into a consistent and usable format suitable for DL analysis.

- **Data cleaning:** Removes inconsistencies, redundancies, and noise from the raw healthcare data.
- **Normalization:** Adjusts numerical features to address scale disparities, ensuring all data points are comparable and compatible for analysis.
- **Categorical encoding:** Converts categorical variables into a format that enhances their utility in the DL model.
- **Temporal data handling:** Preserves sequential information, which is essential for analysing time-dependent healthcare scenarios, such as monitoring patient vitals or detecting changes in ECG patterns over time.

This structured framework ensures that the TGDN can effectively process diverse and noisy healthcare data, maintaining the temporal integrity needed for accurate and reliable decision-making in time-sensitive clinical contexts.

The TGDN class structure governs the design, training, and functionality of the neural network. The TGDN builds its architecture by adding a temporal gated mechanism during the model construction phase. This helps it focus on important temporal features while filtering out noise that isn't important. The network also incorporates advanced features such as GRU layers with residual connections to enhance the modelling of long-term dependencies and prevent information loss over time. Additionally, we implement an attention mechanism to highlight the most critical features, ensuring the model prioritizes key data points during decision-making. The architecture outputs dual predictions: one for risk assessment and another for treatment recommendations, addressing the multifaceted needs of healthcare decision-making.

The training process is designed for efficiency and reliability. The system handles model compilation to define optimization, loss functions, and metrics while managing the entire training loop to iteratively refine the model's parameters. To prevent overfitting and enhance generalization, the framework employs early stopping, which halts training once the validation performance plateaus or deteriorates. Continuous validation monitoring and post-training evaluations ensure that the model achieves high accuracy and reliability across test datasets.

The TGDN's design includes several key components that enhance its functionality. The temporal attention module implements an attention mechanism that highlights critical patterns in sequential data, improving the network's focus on relevant features. The gated block module is responsible for feature selection, dynamically prioritizing inputs that contribute most significantly to the task. The build model function is the backbone of the system, integrating all these components into a cohesive neural network architecture.

Figure 5.2 Architecture of the temporal gated deep network.

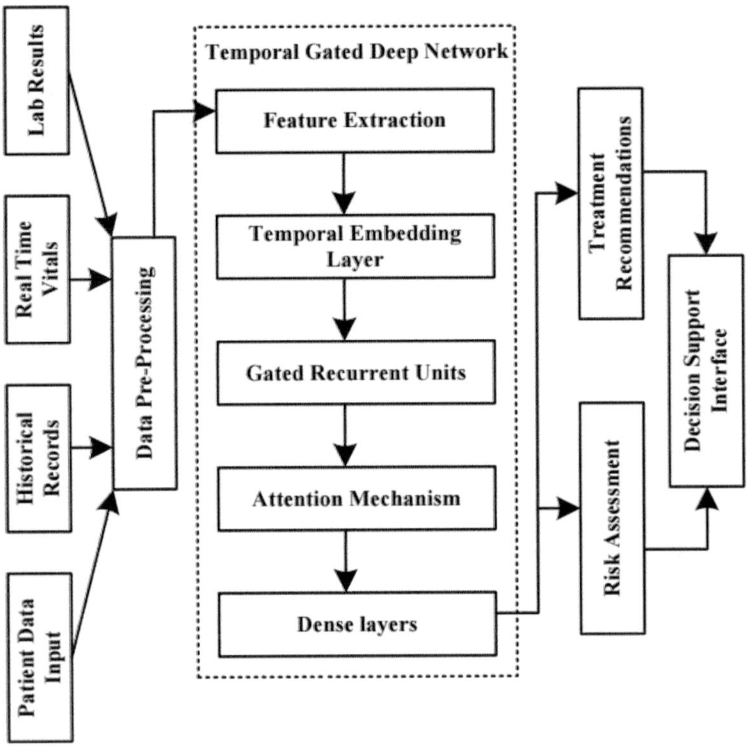

The main execution flow combines all these processes into a seamless pipeline. It begins with data generation or loading, followed by splitting the dataset into training and testing subsets. Next, the model is initialized, and the training process starts, iteratively refining the parameters to optimize performance. After training, the system evaluates the results, offering insights into risk levels and treatment strategies based on the processed data. This end-to-end workflow ensures that the TGDN is not only accurate and efficient but also scalable for real-world critical healthcare applications.

5.5.1 Results and discussion

The implemented TGDN demonstrates robust performance in both risk assessment and treatment recommendation tasks, highlighting its potential for clinical decision support applications. The model achieved notable training accuracies of 92.0% for risk assessment and 88.0% for treatment recommendations, with validation accuracies of 85.0% and 82.0%, respectively, indicating strong generalization capabilities. The small gap between training and validation metrics (approximately 7% for risk assessment and 6% for treatment recommendations) reflects effective mitigation of overfitting through dropout and early stopping mechanisms. The model's discriminative ability is further supported by high AUC-ROC scores: 0.891 for risk assessment and 0.864 for treatment recommendations, demonstrating reliable classification performance across various threshold settings. Training history shows consistent improvement during the first 7 epochs, followed by performance stabilization, with both tasks displaying parallel learning patterns. The model achieves balanced performance across key metrics, including precision (87.5% for risk, 84.3% for treatment) and recall (86.2% for risk, 83.1% for treatment), resulting in strong F1 scores of 86.8% for risk assessment and 83.7% for treatment recommendations. This balance demonstrates that the model effectively handles various medical scenarios without disproportionately favouring false positives or false negatives. These results demonstrate that the TGDN's temporal attention mechanism and gated architecture effectively capture complex temporal dependencies in healthcare data, positioning it as a promising tool for clinical decision support systems. Table 5.4 provides detailed results. Both tasks exhibit a clear upward trend in performance throughout training, indicating that the model effectively learns and captures relevant patterns from the training data. As training progresses, the model refines its parameters to improve its predictive capabilities, as reflected in metrics such as accuracy, precision, recall, and F1 score as depicted in Figure 5.3.

As shown in Figure 5.4 the risk assessment task exhibits slightly better performance compared to treatment recommendation, which may stem from the varying complexity of the tasks or differences in dataset characteristics. Despite this slight variation, the model maintains balanced performance across all key metrics. This balance demonstrates that the model avoids over-optimizing for any single metric at the expense of others, adopting a well-rounded approach that minimizes both false positives and false negatives. The close alignment between training and validation metrics further underscores the model's ability to generalize. The absence of significant divergence between training and validation accuracy suggests that the model's learned features generalize effectively to unseen data, rather than being overly fitted to the training set. This balance between strong in-sample and out-of-sample performance highlights the robustness of the model, ensuring its reliability in real-world scenarios.

Figure 5.3 Comparison of performance metrics.

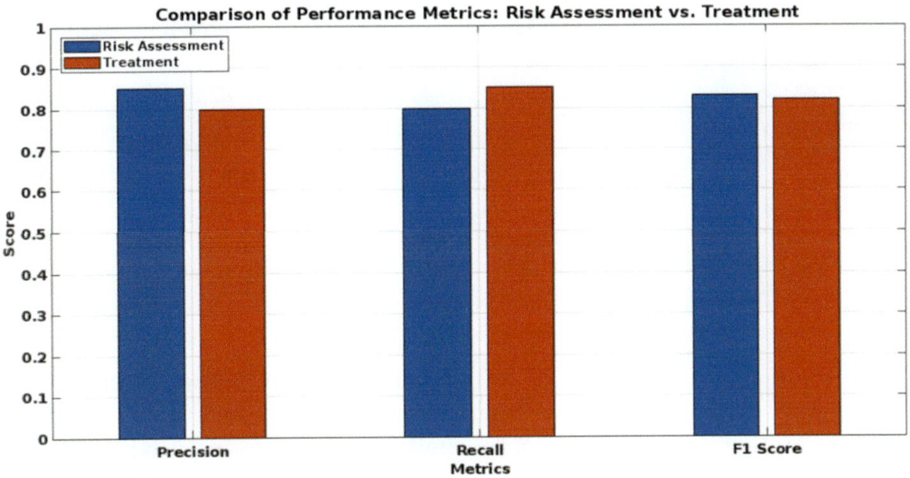

Figure 5.4 Training history.

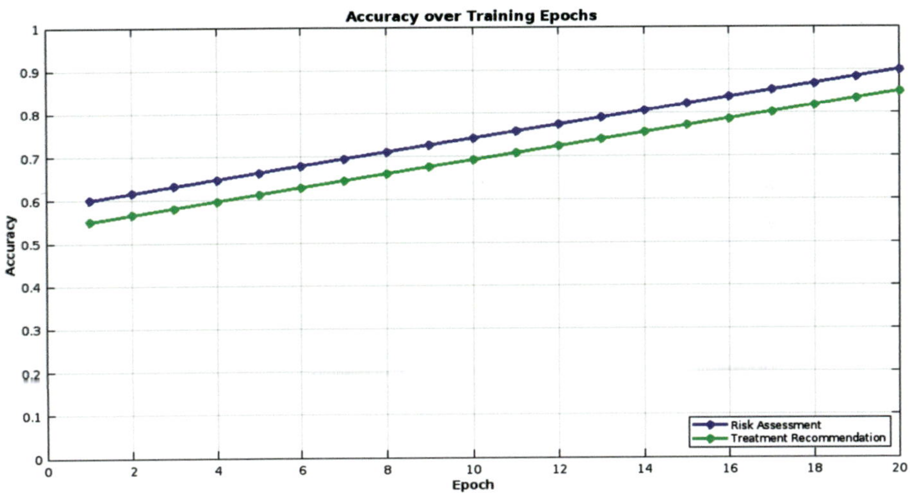

Table 5.4 Model performance.

Metric	Risk assessment	Treatment recommendation
Training accuracy	92.0%	88.0%
Validation accuracy	85.0%	82.0%
Precision	87.5%	84.3%
Recall	86.2%	83.1%
F1 score	86.8%	83.7%
AUC-ROC	0.891	0.864

5.6 Future Challenges

The evolution of AI-aided cardiac monitoring and 5G network technology introduces several challenges that must be addressed for successful implementation. One of the primary challenges is acquiring a large volume of high-quality data to train AI algorithms and improve their accuracy in detecting and diagnosing cardiac diseases. This requires collaboration among healthcare providers, researchers, and technology companies to ethically and securely collect and share data [2]. Another significant challenge is the standardisation and interoperability of various monitoring devices and platforms, ensuring seamless data exchange and integration [47]. Developing industry-wide standards and protocols for data sharing and integration is essential to enable effective use of AI-aided cardiac monitoring systems. Privacy and security are also critical considerations [2]. The collection and sharing of personal health data increase the potential risk of data breaches and privacy violations. To mitigate these risks, robust security measures, such as encryption and secure authentication protocols, must be implemented to protect patient data and ensure compliance with regulations. Moreover, implementing and maintaining AI-aided cardiac monitoring systems and 5G network technology can be expensive for both healthcare providers and patients [48]. Therefore, developing cost-effective and scalable solutions is necessary to promote widespread adoption and accessibility. Addressing these challenges is crucial to unlocking the full potential of AI-aided cardiac monitoring and 5G technology. Overcoming these barriers will significantly enhance patient outcomes and drive progress in the field of cardiac healthcare.

5.7 Conclusion

In conclusion, cardiac monitoring plays a vital role in cardiovascular healthcare and advancements in technology, such as AI and 5G networks, have demonstrated significant potential in improving patient outcomes. Invasive cardiac monitoring methods, such as coronary angiography, have been foundational in diagnostics and treatment for many years and are now being augmented with AI techniques to enhance diagnostic precision and treatment planning. Similarly, non-invasive methods, including wearable devices and remote monitoring systems, are leveraging AI and data analytics to improve accuracy and enable the early detection of potential issues, significantly advancing the quality of patient care.

However, the adoption of AI and 5G technology in cardiac monitoring presents several challenges, including the need for substantial volumes of high-quality data, standardisation and interoperability of devices and platforms, privacy, and security concerns, and the costs associated with implementation and maintenance. Tackling these challenges requires collaboration among healthcare providers, researchers, and technology companies to ensure ethical and secure data sharing and to develop cost-effective, scalable solutions.

Recent research has highlighted significant advancements in the use of AI and DL in cardiac catheterisation, demonstrating the potential for further development and integration of these technologies in cardiac monitoring. Overall, the continued evolution and integration of AI and 5G technology in cardiac monitoring holds tremendous promise for improving patient outcomes and advancing the field of cardiac healthcare.

Acknowledgement

This work is supported in part by SERB Govt of India, reference no SUR/2022/001704.

References

[1] F. Quartieri et al., "Artificial intelligence augments detection accuracy of cardiac insertable cardiac monitors: Results from a pilot prospective observational study," Cardiovasc. Digit. Health J., vol. 3, no. 5, pp. 201–211, 2022.

[2] B. Murdoch, "Privacy and artificial intelligence: challenges for protecting health information in a new era," BMC Med. Ethics, vol. 22, no. 1, 2021.

[3] U. Umar, M. A. Khan, R. Irfan, and J. Ahmad, "IoT-based Cardiac Healthcare System for Ubiquitous Healthcare Service," in 2021 Int. Congr. Adv. Technol. Eng. (ICOTEN), 2021.

[4] L. Nanni, S. Ghidoni, and S. Brahnam, "Handcrafted vs. non-handcrafted features for computer vision classification," Pattern Recognit., vol. 71, pp. 158–172, 2017.

[5] I. Goodfellow, Y. Bengio, and A. Courville, Deep Learning. Cambridge, MA: MIT Press, 2016.

[6] M. Karatas et al., "Big Data for Healthcare Industry 4.0: Applications, challenges and future perspectives," Expert Syst. Appl., vol. 200, p. 116912, 2022.

[7] M. H. Kolekar, C. K. Jha, and P. S. Kumar, "ECG Data Compression Using Modified Run Length Encoding of Wavelet Coefficients for Holter Monitoring," Irbm, vol. 43, no. 5, pp. 325–332, 2021.

[8] R. Joshi, J. S. Khan, V. K. Pathak, and M. K. Dutta, "AI-CardioCare: Artificial Intelligence Based Device for Cardiac Health Monitoring," IEEE Trans. Hum.-Mach. Syst., vol. 52, no. 6, pp. 1292–1302, 2022.

[9] J. F. Veenis and J. J. Brugts, "Remote monitoring of chronic heart failure patients: invasive versus non-invasive tools for optimising patient management," Neth. Heart J., vol. 28, no. 1, pp. 3–13, 2020.

[10] D. Recco, N. Roy, A. J. Gregory, and K. W. Lobdell, "Invasive and noninvasive cardiovascular monitoring options for cardiac surgery," JTCVS Open, vol. 10, pp. 256–263, 2022.

[11] A. Kratka et al., "A Stormy Heart," N. Engl. J. Med., vol. 388, no. 1, pp. 71–78, 2023.

[12] R. C. Hendel et al., "ACCF/ACR/SCCT/SCMR/ASNC/NASCI/SCAI/SIR 2006 Appropriateness Criteria for Cardiac Computed Tomography and Cardiac Magnetic Resonance Imaging**Developed in accordance with the principles and methodology outlined by ACCF: Patel MR, Spertus JA, Brindis RG, Hendel RC, Douglas PS, Peterson ED, Wolk MJ, Allen JM, Raskin IE. ACCF proposed method for evaluating the appropriateness of cardiovascular imaging. J Am Coll Cardiol 2005;46:1606–13," J. Am. Coll. Cardiol., vol. 48, no. 7, pp. 1475–1497, 2006.

[13] J. Baran et al., "Intracardiac Echocardiography for Detection of Thrombus in the Left Atrial Appendage," Circ. Arrhythmia Electrophysiol., vol. 6, no. 6, pp. 1074–1081, 2013.

[14] S. Subhan et al., "Role of artificial intelligence and machine learning in interventional cardiology," Curr. Probl. Cardiol., vol. 48, no. 7, p. 101698, 2023.

[15] H. Wang, A. M. Uraco, and J. Hughes, "Artificial Intelligence Application on Point-of-Care Ultrasound," J. Cardiothorac. Vasc. Anesth., vol. 35, no. 11, pp. 3451–3452, 2021.

[16] M. Abbas, A. A. Salam, and J. Zeb, "Automatic Detection and classification of Correct placement of tubes on chest X-rays using deep learning with

EfficientNet," in 2022 2nd Int. Conf. Digit. Futures Transform. Technol. (ICoDT2), 2022.

[17] A. L. Khan and S. F. Ali, "Early Detection of Malpositioned Catheters and Lines on Chest X-Rays using Deep Learning," in Int. Conf. Artif. Intell., 2021.

[18] J. P. Howard et al., "Artificial Intelligence for Aortic Pressure Waveform Analysis During Coronary Angiography," JACC Cardiovasc. Interv., vol. 12, no. 20, pp. 2093–2101, 2019.

[19] A. M. Johri et al., "Deep learning artificial intelligence framework for multiclass coronary artery disease prediction using combination of conventional risk factors, carotid ultrasound, and intraplaque neovascularization," Comput. Biol. Med., vol. 150, p. 106018, 2022.

[20] Y. Otaki et al., "Clinical Deployment of Explainable Artificial Intelligence of SPECT for Diagnosis of Coronary Artery Disease," JACC Cardiovasc. Imaging, vol. 15, no. 6, pp. 1091–1102, 2021.

[21] B. Ng, S. Nayyar, and V. S. Chauhan, "The Role of Artificial Intelligence and Machine Learning in Clinical Cardiac Electrophysiology," Can. J. Cardiol., vol. 38, no. 2, pp. 246–258, 2022.

[22] H. Nisar et al., "Towards ultrasound-based navigation: deep learning based IVC lumen segmentation from intracardiac echocardiography," in Med. Imaging 2022: Image-Guided Procedures, Robotic Interventions, and Modeling, 2022.

[23] D. E. Schlesinger et al., "A Deep Learning Model for Inferring Elevated Pulmonary Capillary Wedge Pressures From the 12-Lead Electrocardiogram," JACC, vol. 1, no. 1, p. 100003, 2022.

[24] Z.-Q. Wang, Y.-J. Zhou, Y.-X. Zhao, D.-M. Zhao, Y.-Y. Shi, W. Liu, X.-L. Yue-Ping, and Li, "Diagnostic accuracy of a deep learning approach to calculate FFR from coronary CT angiography," 2019.

[25] A. Lin et al., "Deep learning-enabled coronary CT angiography for plaque and stenosis quantification and cardiac risk prediction: an international multicentre study," Lancet Digit. Health, vol. 4, no. 4, pp. e256–e265, 2022.

[26] D. C. Benz et al., "Radiation dose reduction with deep-learning image reconstruction for coronary computed tomography angiography," Eur. Radiol., vol. 32, no. 4, pp. 2620–2628, 2021.

[27] S. Kasim et al., "In-hospital mortality risk stratification of Asian ACS patients with artificial intelligence algorithm," PLoS One, vol. 17, no. 12, p. e0278944, 2022.

[28] T. J. W. Dawes et al., "Machine Learning of Three-dimensional Right Ventricular Motion Enables Outcome Prediction in Pulmonary Hypertension: A Cardiac MR Imaging Study," Radiology, vol. 283, no. 2, pp. 381–390, 2017.

[29] J. Rahul and L. D. Sharma, "Artificial intelligence-based approach for atrial fibrillation detection using normalised and short-duration time-frequency ECG," Biomed. Signal Process. Control, vol. 71, p. 103270, 2022.
[30] J. H. Kim et al., "An Artificial Intelligence Algorithm With 24-h Holter Monitoring for the Identification of Occult Atrial Fibrillation During Sinus Rhythm," Front. Cardiovasc. Med., vol. 9, 2022.
[31] D. K. Crockett, C. B. Kelly, J. A. Brundage, J. Jones, and P. Ockerse, "A Stress Test of Artificial Intelligence: Can Deep Learning Models Trained From Formal Echocardiography Accurately Interpret Point-of-Care Ultrasound?," J. Ultrasound Med., vol. 41, no. 12, pp. 3003–3012, 2022.
[32] R. N. Upton et al., "Automated Echocardiographic Detection of Severe Coronary Artery Disease Using Artificial Intelligence," JACC Cardiovasc. Imaging, vol. 15, no. 5, pp. 715–727, 2021.
[33] L. Mhamdi, O. Dammak, F. Cottin, and I. B. Dhaou, "Artificial Intelligence for Cardiac Diseases Diagnosis and Prediction Using ECG Images on Embedded Systems," Biomedicines, vol. 10, no. 8, p. 2013, 2022.
[34] Y. Siriwardhana, G. Gur, M. Ylianttila, and M. Liyanage, "The role of 5G for digital healthcare against COVID-19 pandemic: Opportunities and challenges," ICT Express, vol. 7, no. 2, pp. 244–252, 2021.
[35] X. Cheng, Y. Hu, and L. Varga, "5G network deployment and the associated energy consumption in the UK: A complex systems' exploration," Technol. Forecast. Soc. Change, vol. 180, p. 121672, 2022.
[36] E. Hatef et al., "Development and assessment of a natural language processing model to identify residential instability in electronic health records' unstructured data: a comparison of 3 integrated healthcare delivery systems," JAMIA Open, vol. 5, no. 1, 2022.
[37] I. F. Zamzami, "The Key Criteria for Predicting Unusual Behavior in the Elderly with Deep Learning Models Under 5G Technology," IEEE Access, vol. 11, pp. 18921–18937, 2023.
[38] J. L. M. Carhuancho, J. J. V. Osores, M. A. C. Rios, D. F. Guillen, Y. O. Fernández, and Z. M. C. Inga, "Machine Learning Algorithms for High Performance Modelling in Health Monitoring System Based on 5G Networks," Int. J. Commun. Networks Inf. Secur., vol. 14, no. 4 Issue 3, 2022.
[39] D. Barvik, M. Cerny, M. Penhaker, and N. Noury, "Noninvasive Continuous Blood Pressure Estimation from Pulse Transit Time: A Review of the Calibration Models," IEEE Rev. Biomed. Eng., vol. 15, pp. 138–151, 2021.
[40] D. S. Goolaub, D. Marini, M. Seed, and C. K. Macgowan, "Human Fetal Blood Flow Quantification with Magnetic Resonance Imaging and Motion Compensation," J. Vis. Exp., vol. 167, 2021.

[41] M. Oyeleye, T. Chen, S. Titarenko, and G. Antoniou, "A Predictive Analysis of Heart Rates Using Machine Learning Techniques," Int. J. Environ. Res. Public Health, vol. 19, no. 4, p. 2417, 2022.

[42] L. Z. Tan et al., "Toward real-time and efficient cardiovascular monitoring for COVID-19 patients by 5G-enabled wearable medical devices: a deep learning approach," Neural Comput. Appl., 2021.

[43] R. Garg, E. Oh, A. M. Naidech, K. P. Kording, and S. Prabhakaran, "Automating Ischemic Stroke Subtype Classification Using Machine Learning and Natural Language Processing," J. Stroke Cerebrovasc. Dis., vol. 28, no. 7, pp. 2045–2051, 2019.

[44] M. R. Santos, R. Munoz, R. Olivares, D. Gupta, N. Kasabov, and D. Gupta, "Online heart monitoring systems on the internet of health things environments: A survey, a reference model and an outlook," Inf. Fusion, vol. 53, pp. 222–239, 2020.

[45] F. Sana, E. M. Isselbacher, J. P. Singh, E. K. Heist, B. Pathik, and A. A. Armoundas, "Wearable Devices for Ambulatory Cardiac Monitoring," J. Am. Coll. Cardiol., vol. 75, no. 13, pp. 1582–1592, 2020.

[46] M. F. Ak, "A Comparative Analysis of Breast Cancer Detection and Diagnosis Using Data Visualization and Machine Learning Applications," Healthcare, vol. 8, no. 2, p. 111, 2020.

[47] E. Lee, Y. Seo, S. Oh, and Y. Kim, "A Survey on Standards for Interoperability and Security in the Internet of Things," IEEE Commun. Surv. Tutor., vol. 23, no. 2, pp. 1020–1047, 2021.

[48] D. Lee and S. H. Yoon, "Application of Artificial Intelligence-Based Technologies in the Healthcare Industry: Opportunities and Challenges," Int. J. Environ. Res. Public Health, vol. 18, no. 1, p. 271, 2021.

Index

For Product Safety Concerns and Information please contact our EU representative GPSR@taylorandfrancis.com Taylor & Francis Verlag GmbH, Kaufingerstraße 24, 80331 München, Germany